TURNAROUND

Musings on the Earth's Future

Tilbury House, Publishers
2 Mechanic Street
Gardiner, Maine 04345
800-582-1899 • www.tilburyhouse.com

Kieve Affective Education, Inc.
PO Box 169
Nobleboro, Maine 04555
207–563–5172 • www.kieve.org

First printing: May 2004 • 10 9 8 7 6 5 4 3 2 1

The quote from "Two Tramps in Mudtime" by Robert Frost is taken from *The Poetry of Robert Frost*, edited by Edward Connery Lathem, published by Holt, Rinehart, and Winston, 1969. Used by permission.

Cataloging-in-Publication Data

Myers, Edward A., 1917-2002.
Turnaround : musings on the earth's future / Edward A. Myers ; edited by Melissa Waterman.— 1st Amer. pbk. ed.
p. cm.
Includes bibliographical references.
ISBN 0-88448-263-4 (pbk. : alk. paper)
1. Global environmental change. 2. Environmentalism. 3. Human ecology. I. Waterman, Melissa. II. Title.
GE149.M94 2004
333.72—dc22

2004001830

Cover painting of Edward A. Myers by his son, Winslow Myers.
Designed on Crummett Mountain by Edith Allard, Somerville, Maine.
Printing and binding by: Maple Vail, Kirkwood, New York. (Printed on Perfection Antique Recycled paper.)
Covers printed by the John P. Pow Company, South Boston.

TURNAROUND

Musings on the Earth's Future

EDWARD A. MYERS

Edited by Melissa Waterman

Tilbury House, Publishers • Gardiner, Maine
Kieve Affective Education, Inc. • Nobleboro, Maine

What shaped the direction of this book
was my husband's fervent wish
for all the world's children
to grow up on a sustainable earth.
To this end, his book is dedicated
to our beloved children and grandchildren.
—Julia B. Myers

CONTENTS

Preface —Richard C. Kennedy *vii*
A Brief Biography *x*
Introduction *xii*

AIR

Thoughts on Energy 3
Rationale for a Sobering Warning and for a Possible Solution 5
Thoughts on Carbon Dioxide 9
Here Comes the Sun 17
Mercator Miasma 25
Climate Change —Peter Shelley 32

NATURE

The Festival of Lights 41
"What Are Your Instructions?" 44
E. F. Schumacher Revisited 49
Ed Myers: Mentor, Investor, Friend —Tom Chappell 54

OCEAN

Nations and Fish: Conflict Resolution on Georges Bank 61
How Big Is Our Garden? 67
Laws and Kings Can't Save Souls 73
The Grim Reaper's Bycatch 76

Man or Schmuck? 79
The Trouble with Bob's Rapeseed 82
Local Action Holds Promise for Sustainable Fisheries —ROBIN ALDEN 85

SPIRIT

The Holy and the World 95
Edward O. Wilson's *The Future of Life* 100
Shepherds and Shepherdesses 104
How to Enjoy an Election Year 108
Marshalling Diversity 115
A Sense of Wonder: Encounters with Ed Myers —W. DONALD HUDSON, JR. 120

Letters 123
Highlights of Progress —MELISSA WATERMAN 145
Bibliography 153

Preface

EDWARD MYERS was a marvelous person, a study in contrasts, and a precious resource for his family, his friends, and the global community. This book is his legacy for all of us; nobody can read this book without thinking in a new and positive way about our chances for survival.

Edward points out innumerable practical ways that you, and I, and, in fact, every individual on the planet, can improve our chances for a long and happy life—and hand on to posterity a far better world than the one we have inherited.

Does he sound like a hand-wringing pessimist or a cockeyed optimist? Neither. He sounds like himself: a realist with an engaging sense of humor who marshals (sometimes) weird facts to support his case. He is the grandfather we all wish we had when we were young. He'd take our questions seriously and offer a soul-satisfying answer when we asked "Why is the sky blue and the grass green?" When you got through with his answer, you would want to look up the words you hadn't understood and you'd feel that your efforts would make a difference.

He was one of a kind, reveled in contrasts and spoke and acted for effect. Nobody forgot Edward Myers—or his message. Who else do you know who rowed a dory on a mussel farm while wearing a Brooks Brothers shirt, a regimental striped bow tie, an old pair of mud-encrusted boots, and a ceaseless smile on his face? He was neither patrician nor peasant, but he felt perfectly comfortable in the company of carpenters and kings. He was enormously articulate but a listener as well. His unorthodox spirituality permitted him to attend church but propelled him to sustain a ministry to longtime prisoners in the state penitentiary. He saw the humor in almost everything that happened but he was deadly serious about saving the planet and turning it over to successive generations in better shape than he found it.

In short, Edward Myers has given us a book of inestimable value. When you read it you will feel stronger and better able to do something positive. Use it first as a primer for yourself; then use it with your family and friends. This book is so full of silver bullets that you will have fun using it to slay the dragons who belong to those who have given up on making our world a healthy place to live.

I wish you had known Edward Myers personally. But the next best thing to that friendship—even kinship—is to know him through his book. You will, I promise, become a far better informed, more dedicated steward of the earth's resources once you have climbed inside the mind of Edward Myers.

RICHARD C. KENNEDY, Founder
Kieve Affective Education, Inc., Nobleboro, Maine

A Brief Biography

EDWARD MYERS, the father of Maine's aquaculture industry, passed away on September 19, 2002. Myers, who was eighty-five, created the first mussel farm in North America near his home in South Bristol, Maine. He also is well known along Maine's coast for his early role in the lobster business, his witty newspaper columns, and his dignified presence—that of an erudite gentleman in glasses and a bow tie.

Myers was also a lay preacher who counseled prison inmates and a peace activist who traveled to Nicaragua in the 1980s.

"He's a major figure because of what he did professionally and who he is personally," said Robin Alden of Stonington, a former Maine commissioner of Marine Resources.

Myers was born in Hillside, New Jersey, and was a graduate of Phillips Exeter Academy and Princeton University. He served in the U.S. Army during World War II, and he and his wife, Julia Booth Myers, settled in Damariscotta, Maine, in 1949.

In the 1950s Myers founded and operated Saltwater Farm and pioneered the business of shipping lobsters around the country. He was administrator of the University of Maine's Darling Marine Center at Walpole from 1969 to 1974.

In 1973 Myers established Abandoned Farm, North America's first mussel cultivation operation. Myers developed modified growing techniques based on work done in Europe. He helped write many of the state aquaculture laws and policies and was granted the state's first aquaculture lease.

"He did aquaculture the hard way—he never made any money at it," Alden said. Myers' work would pave the way for a growing industry that now produces salmon as well as mussels and other shellfish.

"He began the aquaculture business in this state," Alden said. "He's one of those people who had the courage to follow his own paths."

Myers was a student of classic literature, a "man of letters," said Alden. He wrote numerous columns over the years, and most recently penned the philosophical and witty "All at Sea" column published in the Island Institute's *Working Waterfront* newspaper.

He was appointed by several governors to study commissions and oversight committees, including the advisory council to the Maine Marine Resources Department.

Among many civic activities, he was active in area churches and visited weekly with inmates at the Maine State Prison. He was a director of the Maine Peace Mission and a founding director of Citizens Opposing Nuclear Arms, of Damariscotta. He went to Nicaragua twice in the late 1980s, first to promote peace and then to help build a children's day care center.

JOHN RICHARDSON
(Reprinted with permission of *Portland Press Herald/Maine Sunday Telegram.)*

Introduction

WHY DID THIS DOCUMENT come into existence? Why should you read it? And why should you teach it to others?

This book begins in a previous century. My paternal grandfather was born in Philadelphia in 1850. My maternal grandfather was born in Belfast, Maine, in 1853. That was the decade in which the earth's population passed 1 billion, doubling from a half billion in the previous two centuries.

By 2020, 170 years after the arrival of my grandfather, the planetary population will have increased by *eightfold*. Eight billion souls, with bodies to match, are in your immediate future.

During my lifetime, about eighty-five years so far, I have witnessed the doubling of the first billion in 1930, its quadrupling in 1975, and the achievement of 6 billion in October, 1999. I do not expect to be around for the 8 billion.

But you probably will.

This book is not about population control—we are all sufficiently aware of the urge to procreate—but about the burgeoning of a single species, our own, which got its start perhaps 5 million years ago, and the effect that species has upon this planet.

For 99.999 percent of that 20 million years, the earth was in balance. Accumulation of carbon dioxide (CO_2) was absorbed by the forests and land vegetation (28 percent of the earth's area) and by the plankton in the world's oceans (the remaining 72 percent). Oh, there were volcanoes and tidal waves, floods, glaciers, and sea level changes, but nothing so serious that the earth couldn't recover and continue on its merry way.

That is no longer the case.

The earth has outrun its ability to absorb the gases and pollutants thrown at it by us, the 6 billion (and soon enough, the 8 billion). As for CO_2, an easy gas to measure, approximately 4 billion tons no longer have anywhere to go except into the atmosphere. Power plants, automobiles, trucks, factories, and people provide over 20 tons of CO_2 per square mile of the earth every year. And, as you might know, CO_2 is a persistent greenhouse gas which lets in the heat of the sun but traps a good share of it from getting out, and thereby causes climate disruption.

Climate disruption is a term used by Dr. George Woodwell, and is more apt than "global warming." Climate disruption is already starting to melt the Arctic tundra which, possibly to your surprise and certainly to mine, covers nearly one-fifth of the planet's currently dry land. In some cases more than a mile deep, tundra contains the frozen remains of dead animals like lemmings and caribou. As the tundra and its organic cargo melts, another greenhouse gas, methane, will be released into the atmosphere.

No one knows the proportions of the various gases found in the atmosphere 3.5 billion years ago, but we do know earth had a predominately methane atmosphere at that time. Then along came oxygen, deadly to methane-adapted life, which caused a mass extinction of everything except bacteria. Those ingenious bacteria adapted, invented photosynthesis to absorb CO_2, released more oxygen, and 3.25 billion years or so later, here we are.

No one knows the proportions of carbon dioxide in the atmosphere that we and the other animals can tolerate. So we are presently engaged in a gigantic planetary chemical experiment to see how much CO_2 and its global effects we can stand.

When the experiment is complete, there won't be anybody around to report on it. The Creation, in writer Annie Dillard's phrase, will be playing to an empty house.

This dreary prospect does not have to happen.

At the end of the nineteenth century, we made a choice or

rather, had it made for us. We would run the world on fossil fuels. It proved to be a very successful choice for a runaway industrial society and worked well, in its way, for about a century. But for the long-term the choice was, in a word, lousy. Now we have a chance to turn it around. Our only chance. "God's Last Offer," as Ed Ayres said in his eponymous book.

Listen to a monk, Thomas Berry:

> The Great Work before us, the task of moving modern industrial civilization from its devastating influence on the earth to a more benign mode of presence is not a role we have chosen. It is a role given to us beyond any consultation with ourselves. The nobility of our lives, however, depends upon the manner in which we come to understand and fulfill our assigned role.

Can you imagine a more exciting assignment?

If this assignment turns out to be a life-and-death struggle, as it might, then our task *has* to be done. If it turns out to be just a thorough scare, it is still worth doing. Your actions could clean up the planet's air. The enormous increase in respiratory diseases will be reversed, among other benefits. The planet's drinking water would again be plentiful, relieving the thirst of today's 3 billion people who do not have access to potable water. The world's oceans will be clean and again teeming with fish, the opposite of the present situation in which entire fish stocks collapse, the sea level rises disastrously, and unseen but deadly threats from persistent organic pollutants accumulate in the marine food chain.

"But," you say, "won't this hurtle us back into the seventeenth century, to the world as it was before the Industrial Revolution?"

No way. Absolutely not.

The Industrial Revolution, for all its faults and Faustian bargains, resulted in trillions and trillions of dollars added into the world. Another trillion's worth of gross world product is added every few years (although the Third World hasn't seen too much of that yet, it's on its way). In the U.S. alone, between now and 2020, at least 12 trillion dollars of wealth will be bequeathed to the next generation by people over the age of seventy.

With those funds applied properly, job growth and fascinating professions will follow. Designs are in the works today for zero-emission factories, automobile assembly plants, and power plants requiring the skills of designers, architects, engineers of all stripes, chemists, physicists, and many others. If the U.S. needs 20 million new housing units by 2020, as predicted, skillful planning and design could meet that goal and still leave 2.5 million acres in open space, making cities livable, eliminating expensive sprawl—and someone would still make lots of money.

It's a bright prospect. The list of things that need to be done is endless, the challenge is marvelous, and the outcome will be good for the whole planet.

Through it all, keep in mind that all your decisions should have a spiritual base. We owe that to ourselves, our children, and our grandchildren, all those that follow, and to the Creation, however you may understand that.

Separate your objectives and your spirituality from your religion. There are hundreds of religions on the planet, each suggesting a slightly different way to pursue happiness, achieve mental comfort, and minimize suffering. Some religions historically have fought savagely to convince others that theirs is the only true way. It is too late for that. Instead, be on an absolutely practical planetary quest, possibly for the survival of all life, and certainly for good air to breathe, good water to drink, and enough nourishment for everybody on earth, plus all the animal species we have left or can restore. The quest won't solve all the earth's problems, but is definitely something that will demand all the spirituality the world can assemble, starting with you.

The fact that it probably will be genuine fun, adequately profitable, and solidly sustainable is just icing on the cake.

Oh, and those grandfathers mentioned on the first page? Both were transferred to New York City by their companies. Both looked for quiet places to raise their families, places that would never change. One chose a nice Cape with a white picket fence in Bayonne, New Jersey. Now there's an Exxon storage tank where the house used to be. The other bought a farm in Hillside, New Jersey. The town is now subsumed in the sprawl of Elizabeth and Newark, its fields covered with houses every fifty feet, and jets taking off every twenty-six seconds from the airport runways three miles distant. Let this be a lesson to you.

ONE

AIR

Thoughts on Energy

BEFORE WE START, let's get located. Forget all the buildings, pavement, supertankers, and the rest of that stuff. You are on the Good Earth.

How do we know it's good? Because for about five billion years, this 6 sextillion, 5 quintillion-ton, not-quite-round ball has been hurtling around the sun at 66,600 mph, spinning all the while at 21,600 miles per day in a solar system itself moving 15,000 mph somewhere else in space. It has no visible means of support, no brakes, but it hasn't collided with anything except some asteroids now and then, millions of years apart. And even then, it has recovered.

Furthermore, it has an atmosphere, essential to all life, and heavy enough (about fifteen pounds per square inch at the earth's surface) to ride along without being stripped away, thank heaven.

Now we get into some big numbers which are hard to grasp, and eventually involve some guesswork. There's no need to remember them exactly, they are simply intended to give a picture. At that fifteen pounds of pressure per square inch, the atmosphere weighs 612 trillion tons. A big number, yet still only one-billionth of the planet's weight.

The atmosphere is what keeps all life alive. It's a miracle, the miracle of miracles. Three or so million millennia ago, Somebody or Some Process decreed: "This breathing of methane has no future. Let's start over. From now on it's going to be 78 percent nitrogen, 21 percent oxygen, and 1 percent argon, plus a few trace elements of this and that. Let's see if this mix won't work!"

And so it has, with marvelous results as far as we are concerned.

Yet there's something wrong. Instead of looking up to the sun as a power source, which operated all this, we looked down, into the earth for the finite and non-renewable stored power of the sun—for veins of coal, pools of oil, pockets of natural gas, and, for the last sixty years, uranium.

Even Thomas Edison, the Wizard of Menlo Park, holder of a thousand patents largely in the generation of power and light, founder of Edison Electric Light Company in 1878, and inventor of moving pictures in 1887, expressed the wish a few years before he died in 1931 that the world would revert to harnessing the rays of the sun and its earthly handmaidens, the wind and the rain, rather than rushing headlong into the further use of fossil fuels.

Nobody listened to Tom Edison. Now it's time to.

Rationale for a Sobering Warning and for a Possible Solution

Sermon, 30 June 2000

THE UNITED STATES OF AMERICA, now at 260-some million humans and working toward 400 million by 2025, is headed for some trouble. The world, at 6 billion as of October 1999, is working toward 8 or 9 billion by 2020, and is also headed for some trouble. No one can predict the extent of it. About 98 percent of the people are not paying attention or haven't a clue about what is coming or simply deny that any trouble is coming.

Any or all of those opinions held by a vast majority are unwise.

The earth is engaged in a gigantic chemistry experiment: How many tons of carbon particulates, carbon dioxide, sulphur dioxide, dioxin, chlorofluorocarbons, polychloride biphenyls, nitrous oxide, and many other pollutants can be presented to the earth's atmosphere before the entire living Creation is suffocated?

If the people wait for the final answer, the answer will be very final indeed.

As T. H. Huxley said,

> The chessboard is the world; the pieces are the phenomena of the universe; the rules of the game are what we call the laws of Nature. The player on the other side is hidden from us. We know that his play is fair, just, and patient. But also we know, to our cost, that he never overlooks a mistake, or makes the smallest allowance for ignorance.

With this warning in mind, we know from the paleontologists and the geologists that a self-healing and self-correcting planet goes through long cycles. Perhaps 3 billion years ago, the liquid of the world ocean became a saturated solution of carbon dioxide. What life there was, probably microbial, was suddenly presented with oxygen, a deadly poison gas, to which they adapted, invented photosynthesis, and eventually made humankind possible.

A smaller example of the self-correcting strength of the earth is the 1815 eruption of Tambora in the East Indies, which produced worldwide dust clouds, resulting in snowstorms and killing frosts in every month of 1816 in the northern hemisphere. But again, the earth corrected the problem.

Two major sinks preserve breathable air from destruction by carbon: the world's oceans and the world's forests. In ecology a sink is a receptacle for certain elements of the system. No one knows how much carbon these two sinks can absorb. Both are beset.

In the ocean plankton are the basic photosynthesizing biomass, absorbing CO_2 and releasing oxygen. The plankton may be doing their best, but they are weakened by the effects of the ozone hole, which cannot be repaired before mid-century; by the melting of the Ross and Larsen B ice sheets; by persistent organic pollutants; by harmful algal blooms; by oil spills; and by invasive species transferred from ships' bilges. On land trees absorb CO_2 yet they are being cleared at the rate of 47,000 acres per day, a rate which, if not increased, will reach the world's last tree in 2101. These two sinks are currently being asked to absorb 6.75 billion tons of CO_2 per year.

The U.S. won't sign the Kyoto Protocol. Same for Russia and China. The governments talk about lack of political will and are mostly irrelevant, particularly the U.S. Congress. In 1992 former President Clinton promised to reduce greenhouse gases by 15 percent before 2000; in the ensuing eight years, U.S. total emissions of $C0_2$ have grown about 20 percent. The primary increase has come from SUVs and light trucks, both exempted from pollution controls by Congress. China says it will build 5 million vehicles per year beginning in 2005. Powering the manufacture will be soft coal, creating a major leap in greenhouse gases. India has passed 1 billion humans in population and so will continue its position as a top polluter.

Whether or not the world is proceeding toward a horrible catastrophe, there is now a magnificent opportunity to leave the industrial revolution-fossil fuel era behind in favor of renewable energy, pollution free. There are good indications that we have begun:

- Enough sunlight reliably falls on the earth from 10 degrees S. to 10 degrees N. to supply the energy needed for all the world.
- By report, there is sufficient wind power on American Indian reservations in the U.S. to supply the whole country.
- Wind power produced 13,840 megawatts of electricity worldwide in 1999; in 1980 it was 10 mgw. Germany leads in production; the U.S.A. is second, then Spain, Denmark, Italy, and Greece.
- Power from photovoltaic modules grew from 0.1 mgw in 1970 to 201 mgw in 1999. [Their price] fell from $5.50 per watt in 1987 to $3.50 per watt in 1999. Royal Dutch Shell operates a 25-mgw cell factory in Germany.
- Honda and Toyota are marketing electric–fossil fuel hybrid cars now. A Honda hybrid traveled from South Boston to Wiscasset on three gallons of gas—approximately the 70 percent reduction of greenhouse gases suggested at Kyoto—averaging 55 mph door-to-door.
- Daimler Chrysler has chosen Iceland as its test area for a fuel cell/electric car. The fuel is electrolyzed hydrogen; the only exhaust is pure water.
- William McDonough Associates has a $2-billion budget from Ford to turn its River Rouge automobile plant into a zero-emissions plant.

The above are a few indications that the next quarter century can deliver the most exciting transition from a dangerously polluting society to one in harmony with the planet.

There is little to worry about economically—the transition can provide enough work for everybody in hundreds of different fields. The 1999 installations for wind power alone supported the creation of 86,000 jobs.

The most important and most difficult change will be the embedding of the idea of total energy transition in the minds and hearts of the populace. No reliance can be placed in elected and other governments until the people, now lulled by prosperity, inertia, and denial, provide the political will to urge governments to cooperate with one another in a crisis that is without boundaries.

It no longer matters whether or when the world will run low on petroleum, or coal, or any other fossil fuel. What matters is making the transition to the natural power of the sun as soon as possible.

Thoughts on Carbon Dioxide

I'LL BEGIN WITH THE END of the late Donella Meadows's last column. Each sentence packs a wallop, especially the last one:

> The economic rule is: Do whatever makes sense in monetary terms.
>
> The Earth says: Money measures nothing more than the relative power of some humans over other humans, and that power is puny, compared with the power of the climate, the oceans, [and] the uncounted multitudes of one-celled organisms that created the atmosphere, that recycle the waste, that have lasted for 3 billion years. The fact that the economy, which has lasted maybe 200 years, puts zero value on these things means only that the economy knows nothing about value—or about lasting.
>
> Economics says: Worry, struggle, be dissatisfied. The permanent condition of humankind is scarcity. The only way out of scarcity is to accumulate and hoard, though that means, regrettably, that others will have less. Too bad, but there is not enough to go around.
>
> The Earth says: Rejoice! You have been born into a world of self-maintaining abundance and incredible beauty. Feel it, taste it, be amazed by it. If you stop your struggle and lift your eyes long enough to see Earth's wonders, to play and dance with the glories around you, you will discover what you really need. It isn't that much. There is enough. As long as you control your numbers, there will be enough for everyone and for as long as you can imagine.

> We don't get to choose which laws, those of the economy or those of the Earth, will ultimately prevail. We can choose which ones we will personally live under—and whether to make our economic laws consistent with planetary ones, or to find out what happens if we don't.

What Meadows is saying is a question asked for centuries. It comes when you look in a mirror and ask: "If not me, who? If not now, when?"

So let us begin with the Number One Global Problem: the buildup in the atmosphere of 4 billion tons of carbon dioxide (CO_2) every year. One country (guess who?), with a shade over 4 percent of the world population, contributes one-quarter of that total. That's right—a billion tons comes from the U.S.A., the super-user of fossil fuels.

The U.S.A. has a love affair with the automobile, and no other country comes close in exhaust emissions. The country is also fond of electric power, and no other country comes close to that, either. Compare U.S. figures with those of Africa's largest nation, Sudan. Sudan has 35 million people; the U.S. has 275 million. The Gross Domestic Product (GDP) in Sudan per person per year is $930; the GDP in the U.S. is $31,500. There are 75,000 cars and trucks in Sudan, about one vehicle for every 466 people. In the U.S. there are 206 million cars and trucks, or one vehicle for every 1.3 people. Sudan uses 1.8 billion kilowatt hours of electricity each year; the U.S. uses 3,620 billion kilowatt hours in a year. Now, don't let that Gross Domestic Product figure fool you. GDP isn't much good as a measure. Everything is tossed into it. If you have a bad car accident, it increases the GDP, because of the work done by hospitals, doctors, tow trucks, insurance companies, etc. Its only true use is as a comparison.

Sudan wasn't selected because it is a basket case; it isn't. It's just one of those Third World countries, as we describe them.

In Sudan 35 million people rise each morning and go about their daily chores on less than 3 percent of the U.S. GDP, fewer than 0.003 percent of U.S. vehicles, and 0.05 percent of electric power.

The northern hemisphere, in which Sudan is located, is home to the nine major CO_2 polluting countries which, together, provide more than 70 percent of CO_2 causing climate disruption. A lot of it rides the jet stream and is carried toward the North Pole first, but CO_2 does stay in circulation. It eventually makes its way around the world, so that Sudan and every other country gets a share.

To get a notion of what just one poorly designed and sloppily operated power plant can do, consider this study on the Brayton Point Power Station from the Harvard School of Public Health. The station is in Fall River, in southern Massachusetts, and most of the state of Rhode Island is affected by its toxic plume. That plume covers 1,451 square miles. The entire state of Rhode Island is just 1,231 square miles. According to a study conducted by Harvard University, the effluent from Brayton Point causes a hundred premature deaths each year in the state, as well as 30,900 asthma attacks. Your risk of dying is three times greater if you live within a thirty-mile radius of this noxious power plant.

What we learn from these two examples, one local, one international, is to ask the big question: *Quo warranto? By what authority?* By what authority has someone or some corporation or some country been granted the right to kill people, send thousands into the throes of asthma, and trifle with the life spans of many thousands more?

Quo warranto indeed. If you're feeling religious, take a quick run through the Pentateuch with the Ten Commandments, the Sermon on the Mount in the Book of Matthew, the Torah, the Koran, the Upanishads, the Sayings of Buddha, the Analects of Confucious, the Tao Te Ching, and any other creeds you'd like to parse, and then ask yourself how many of them authorize killing at random, tripling anyone's chances of having a short life, afflicting thousands with a respiratory disease that won't go away, and continuing this year in and year out. The answer is, none.

While you're at it, check the Bill of Rights, the first ten amendments of our Constitution. It protects us from cruel and unusual punishments, such as a hundred arbitrary deaths from airborne pollutants, and the denial of life and liberty without due process, and unreasonable searches and seizures. How can this be happening?

Well, it does, because people want electricity. Taking the four coal-fired generators at Brayton Point as an example, what do people get with their electricity?

- 7.93 million tons of CO_2
- 34,300 tons of sulfur dioxide
- 7,120 tons of NO, and various other oxides of nitrogen
- A billion gallons of chlorinated seawater discharged each day into Mt. Hope Bay at a temperature of 95°F
- Arsenic, selenium, vanadium, and more dumped into unlined pits
- Lots of mercury

Which brings up a number of questions. The first is yours: What's this got to do with Sudan?

This one Massachusetts plant, one of thousands in the United States, uses coal to produce annually about as much pollution as that produced by the entire nation of Germany. And it creates about as much power as Sudan will use in a year by 2005.

Keep in mind that we have 466 times more cars and trucks than Sudan does. You can understand why Sudan is just an example of the 175 or so other nations of human beings who, like us, want the "good things" in life. Like us, they are hoping for a long happy life, a minimum of woe, and more of the same for those who come after them.

To achieve that for everyone, something has to be done

before we sag swiftly into great difficulties or are overtaken by atmospheric catastrophe. Whichever occurs doesn't make much difference; we can't go on as we are.

What is being proposed quite simply (and it *is* simple) is a conversion of the whole planet's energy program to the way it was supposed to be. As Donella Meadows said, "You have been born into a world of self-maintaining abundance and incredible beauty.... As long as you control your numbers, there will be enough for everyone and for as long as you can imagine."

Your first reaction is undoubtedly, "Oh, come on. We're going to rewire and refuel the whole planet? Without fossil fuels? You've gotta be out of your mind!"

Nope.

And the reason that can be said is that it has already been done. The wrong way, to be sure, using coal and oil to drive our civilization, but IT HAS ALREADY BEEN DONE.

Cast your mind back, if you will, to 1895. Electricity was in its infancy; the principal measure of horsepower was still the horse; a Stanley Steamer car was a curiosity, forever getting stuck in the mud and frightening the horses. If a cellar was needed, you hired men with shovels. Less than one-fourth of 1 percent of American males attended college; women wouldn't even be allowed to vote for another twenty-five years. The United States population had reached 70 million. The world's population was perhaps 1.5 billion. Farmers made up more than 50 percent of U.S. workers. Thinking that Pennsylvania crude oil would never amount to anything, Rudolf Diesel was trying to find the ideal vegetable oil for his newly invented combustion engine.

And then—Henry Ford, the Wright brothers, Louis Bleriot, Marconi, telephones, radio broadcasting, the transatlantic cable, paved roads, evolving computers—an endless flow of ingenuity and inventions which changed the world forever.

Between 1895 and 1935 the world was set on its path for the rest of the twentieth century. All that inventing got done without much education or preparation or infrastructure and was interrupted by a world war and a major world depression. So it ought to be possible to move onto a new path in twenty years this time, and to do it right.

Is anything being done? You bet.

Two governments have voted to become CO_2-emissions-free by the 2020s. In one case this decision came from the oldest parliament in the world, Iceland. Yes, the country does have a lot of geothermal energy, but it also has to import every drop of fossil fuel for its automobiles, fishing fleet, aluminum smelters, and seaweed processing plants.

The other country is tiny Vanuatu, an independent republic that once was known as the New Hebrides, and was governed by Great Britain and France until 1980. The little island nation is rightly fearful of being drowned by sea-level rise caused by climate disruption as the ocean warms up and expands. One of Vanuatu's smaller neighbors has already applied for asylum in Australia, as its islands are being inundated.

So now we have just 180-odd nations to go. A large number of them are not odd at all. The Netherlands knows what's coming and is developing a master plan to raise the dikes. Germany and India are currently the world's leaders in wind power, although others, including the state of Wisconsin, are catching up. China, with vast veins of soft coal, is nonetheless changing over to natural gas to reduce emissions so that the citizens of Beijing one day may not have to go outside wearing face masks because of the city's air quality. At the end of 1997 there was an ecological summit in Kyoto, Japan. With exceptions countable on the fingers of one hand, every nation signed on to the Kyoto Protocols. No nation questioned the science of them, except one: the United States of America. Of course, and alas.

"Cognitive dissonance" is an easily translatable phrase. You know something that the other fellow doesn't, or he isn't paying attention, or he can't wait to disagree with you because he wants something else, or all three.

Cognitive dissonance is what we end up with in Washington. With all that's going badly in the only atmosphere we have, the U.S. government spent 2001 fuming over election reform, faith-based charity, Social Security (which any good actuary with true data could solve in an afternoon), drilling in the Arctic for oil that might last six months, abortion, and other items. Meanwhile, a fossil-fueled economy, which they think is going to go on forever, shudders and creeps into the future. And no one heeds the fact that no economy is worth anything if there isn't an environment in which you can breathe.

The United States is the most powerful country in the world, the richest country in the world, with the best communications in the world, the most technology in the world, and without a single credible enemy in the world. And does it have the political will to solve the critical problems shadowing us today? No.

So the people have to do it. If the government won't do something vital, the people in a democracy will have to.

The way to begin is to begin. You will be surprised at how much help you will get and how many people will join you.

I'm a little old to be storming the barricades, so instead, I bought a car. My own government wouldn't ratify the Kyoto Protocols, so I did. I bought a little gas-electric hybrid vehicle, one that gets 70.2 miles to the gallon and reduces emissions by 80 percent compared to a standard car. The gas mileage is 2.5 times that of the average U.S. car and the emissions reduction is way ahead of what was called for at Kyoto. Furthermore, the building in which I write to you is well insulated and draws 90 percent of its power from a small wind generator and a photovoltaic panel.

You don't have to do anything quite so drastic. You simply have to prepare yourself for a useful career that moves the world along the right track. And remember:

By choice or catastrophe, civilization will discover a sustainable way of life, and working for an enticing future is better than blundering into a catastrophe.

Here Comes the Sun

Unitarian Universalist Sermon, 9 June 1996

IT IS ALMOST THE END of the school year, and it's time for a review class on "The Planet According to Me." Some of this you may have heard before. Just please forgive an elderly fanatic who has had for nearly eight decades, in Robert Frost's words, "a lover's quarrel with the world."

First, the planet. It weighs 6.588 sextillion tons—that's six and twenty-one zeroes, and to be precise, that's short tons, not metric—and it sits completely unsupported in empty space, held in relationship to other heavenly bodies by forces we can describe but not explain: gravity, delicate balances, and mutual tensions. Well, no, it doesn't sit—it's actually traveling through space at 66,600 miles an hour around the sun, and it's spinning—a fraction of a second slower each year—at 900 nautical miles an hour at the equator and 460 mph at 44 degrees North latitude, which is about where we are; the 44th parallel goes right through the Edgecomb post office. Besides those two speeds—one orbital and one spinning—the whole solar system is hurtling, apparently toward the Spiral Nebula of Andromeda, at some unimaginable rate. We hope it has a long way to go, since running into something else probably wouldn't be safe, not with six sextillion tons of momentum at 18.5 miles per second.

Having absolutely nothing around us except our moon for millions of miles is pretty scary, but it's also absolutely necessary. If you recall my grapefruit metaphor, where I take the earth down to the scale of a hundred-millimeter grapefruit (about four inches), the life-supporting atmosphere—from the deepest point in the ocean to the highest point of the stratosphere—is just half a millimeter thick, a very thin veneer. The miracle is that, no matter how fast we go, the fact that we are going through absolutely nothing means that nothing can blow away the atmosphere. And so, thank heaven, it goes along with us.

Next, time. Taking the best guess, it's been perhaps 14 billion years since the Big Bang or whatever cosmological theory you wish to espouse. Let's take a figure of 4.5 billion for the age of our planet. According to astronomers, 4.5 billion years is about the halfway mark in the life of the sun. The sun is a continuous thermonuclear explosion at the rate of something like 440,000 tons of gas per second and, at 900,000 miles in diameter with the corona, it ought to last another 4.5 billion years before it goes out.

Ninety-nine point nine percent of those first 4.5 billion years passed before hominids appeared, let's say around 4.5 million years ago. Since then we have progressed from *Pithecanthropus erectus* to *Homo sapiens*, meaning *knowing*, and currently, with possibly dubious precision, to *Homo sapiens sapiens*, which means the *man who knows he knows,* a very wise man indeed.

Down again an order of magnitude, to 4,500 years ago, when people worshipped what they didn't understand. As understanding dispelled superstition, there appeared almost simultaneously the notion of a monotheistic religion, and rising to the occasion there appeared Moses and Gautama Buddha and Jesus and Mohammed. This period continues, but it is well to remember that it is within the last fraction of life's time on the earth so far.

Before the agrarian revolution, when humankind left the hunter-gatherer tradition to tend regular crops and domestic animals, the world could rely for its infinitely renewable resources on the power of the sun and the sun's planetary by-product, rain. For most of the last millionth of the earth's history, there was oat power: you fed oat to horses and water buffaloes and elephants and oxen, and the by-products were valuable—manure to enrich the fields and meat and sometimes ivory.

Then, in 1697, Thomas Savery improved on Edward Somerset's 1627 crude attempt at a steam engine. Savery developed what was called "the miner's friend" to pump water out of coal pits. Newcomen and James Watt followed in his footsteps, and the world began to go from oat power to horsepower (which isn't the power of a horse at all but an arbitrary measure of the energy sufficient to lift 33,000 pounds one foot in one minute). James Watt—much can be made of a Scotsman if you catch him young—learned something about condensation and compounding, and from there on it was all downhill, for we traded oats for soft coal and then for other fossil fuels, in which I include uranium isotopes.

Where are we today? We are putting 6 billion tons of carbon particulates into the air and generating uncounted tons of nuclear waste. Nearly 75 percent of carbon particulates come from nine countries—one guess who's number one—and all those countries lie in the northern hemisphere. As the carbon sinks to earth, we must rely on the plants and forests to photosynthesize the CO_2 and release in exchange the oxygen we need to breathe, yet we are paving over the plants and cutting down the forests just as fast as we can. As they sink into the world ocean, we must rely on the plankton to use up the CO_2 and release oxygen. But there again there are limits. We insult the oceans with everything that runs off the land. No one knows how much the plankton can absorb, though we do know the ocean's life is under stress.

I decline to discuss population problems, except that it is perfectly well known that the increase of carbon dioxide in the atmosphere and radioactive waste on the land is linear with population growth. As I have mentioned before, Clinton promised to return the U.S. to its 1990 emissions levels by the year 2000. The actual increase of 57 million tons of CO_2 in the U.S. during the 1990s follows our increase in population.

Finally, the Greenhouse Effect. You can pay a think-tank

scientist to say that it isn't happening. But it is. What a greenhouse does is accept short wavelengths of light and refract long wavelengths of light, which can't get back through the glass. The glass for the atmosphere is the carbon dioxide blanket, and sunlight reflected from the earth's surface can't get through it. The warming effect increases toward the poles, so we should feel it first here. Besides an increase in sea level as the ocean warms and the poles melt, the ultimate irony would be melting the Arctic tundra. Melting tundra and permafrost will release enough methane (C_6H_6) to put the planet back where it was at the beginning, when the earth featured a methane atmosphere with no polar icecaps and no land.

Well, then. For 4,499,999,800 years the earth lived in balance with the sun. Oh, there have been earthquakes and volcanoes, the disappearance of the dinosaurs, tidal waves, and so on, about what you should expect with plate tectonics and a molten core, but it has always returned to balance, except for the last two hundred years, when the domination of humankind has upset the balance. We seem not to want to stop.

How much time do we have to correct the situation? And what do we have to do?

As the local Cassandra without portfolio, I suggest that we have no more than twenty years, and that we had better begin right now. What to do? We return to the sun. We abandon fossil and nuclear fuels. We use all the sun's products—solar heat, hydropower, tidal power, the wind, and geothermal sources.

I am not a Luddite wishing for a return to a romanticized past with cute villages and no machinery. We need a basic clean fuel, one that's provided by the sun. That, of course, is hydrogen. It's the element most abundant in the universe; it burns cleanly; its only by-product is pure water—oh, there's a 0.000156 fraction of deuterium—but you could run a bus on hydrogen all day and have much less of that than the carbon monoxide in one cigarette. And the handiest raw material for hydrogen is plain water, H_20.

Hydrogen is as safe as gasoline, has none of the bad traits, and is easy to distribute by pipeline. Electricity is needed to electrolyze water into hydrogen and oxygen. And with new, efficient wind turbines there is enough wind power in the Dakotas alone to provide all the electricity needed to make all the hydrogen we need to run the whole country. It's a wonderful thought to think of the American Indian reservations in the West not as gambling casinos, but as suppliers of a natural, inexhaustible fuel from the sun for all of us. Think of zero-emission factories—they're possible. Zero-emission power plants and transportation—no more carbon particulates at all. Zero-emission houses, with no further need for septic systems, and with lots of vegetables, growing without season.

Things are beginning right now. There is a very intense Russian hydrodynamicist at MIT who has devised a turbine now operating on the bottom of the Cape Cod Canal whenever the tide is above three knots, at a cost of around 2.5 cents a kilowatt. He also has a land version, which he places on the median strip of Route 128 and the Mass. Pike. The whooshing by of cars and trucks creates the wind power to generate electricity at less than 2.5 cents. The numbers this man runs indicate that such machines could provide all the power needed in New England. The engaging part of this Russian MIT device is that it is shaped like the double helix, the basic shape of our own fundamental building block, DNA.

A half-dozen of these turbines unobtrusively anchored on the floor of the Lower Narrows on the Damariscotta River (three to six knots) and under the bridge at Damariscotta–Newcastle (four to eight knots, maybe nine when the sun, moon, and earth are in syzygy) can supply power for all of Lincoln County. The neat part of this pair of narrows is the ninety-minute time-lag between South Bristol and East Boothbay and the towns' bridge: slack water at one is fast water at the other. Should there be a little hiatus on the neap tides, we can stick a spare turbine in at Glidden Ledge. Then we can take care of Sagadahoc County at the two Hell Gates on the Sasanoa, with the Narrows at the northerly end of Westport for the spare turbine.

Unavoidably, the next step is minimizing the use of nuclear power. Incredibly, it turns out to be easy and economical, and what's more, it's already being done. I asked a twelve-year-old boy at South Bristol School to research energy sources on the Internet, even though I had none of those mysterious addresses. He quickly found a completely self-contained photovoltaic house in Biddeford Pool, which led to an architectural firm in Harvard, Massachusetts, which had a subsidiary that converted the Rancho Seco nuclear plant in Sacramento, California. Up on the screen came a photograph of two giant cooling towers, all cobwebs and dust, and every other inch of the 640 acres which a nuke is required to own covered with solar cells. What had happened?

Being California, its citizens initiated and passed an initiative to form a Municipal Utility District. The district elected citizen directors. At their first meeting, the directors asked Rancho Seco for its operating statements. At the second meeting, after looking at the figures, the board voted to close the plant. At the third meeting, having worked it over, they hired S. David Freeman, who had already closed four nuclear plants while working for the T.V.A., and asked him to convert the plant. The photovoltaic array on Rancho Seco's land came up fifty megawatts short of what the nuclear plant had generated, so they added fifty megawatts of wind power on the top slopes of the valley.

So now the wind and photovoltaic combo is producing just as much power at about half the real cost of nuclear power, with no side effects, no unplanned releases, no refueling, AND no dismantling. Instead of a million dollars a megawatt to take the plant down, they left it there, first shipping off the really hot stuff to Hanford and other intractable dumps. The only penance is to wait sixty years with periodic scrubbing down until the buildings are usable. I'm told that the reactor dome, in the 2050s, can become an acoustically marvelous center for the performing arts.

Can the same procedure be applied to Maine Yankee? Yes, indeed, and the timing is perfect, since the power isn't needed and can be replaced with 2.5-cents-power instead of the unsubsidized real cost of close to 7 cents today. Maine Yankee owns 800 acres of substantially cleared land, and 640 of it can be used for solar panels. At 4,356 solar panels per acre, we might need 570 acres, for 2,482,920 panels. If this notion spread to other plants, such a large order suggests the possibility of a cost around $10 per panel, say $25 million, or just 12 percent of the 1972 cost of the plant it replaces and only 4 percent of the proposed dismantling cost, which isn't necessary. If you need more space, there's cleared land under the miles of transmission line—100 feet wide gives you an acre every 436 feet.

Don't be fooled by our boreal climate. Maine has a few more daylight hours than Tampa, and just about as much sunlight. There are plenty of Maine houses already off the grid or selling power back to a company or a neighbor. Rancho Seco made this the foundation of a new business; the Municipal Utility District is retrofitting hundred of houses and factories each year with photovoltaics, creating jobs for roofers, electricians, and plumbers. Rancho Seco uses a new triple-switch photovoltaic unit (37 cents in 1993, 10 cents in 1994) that allows them to electrolyze hydrogen to fuel a steam-driven plant needed when the sun is hidden by thick clouds. The district's goal is to eliminate the need to construct the next power plant on the drawing board.

Well, that's the vision. It's big but it's necessary. It involves a complete changeover of power for cars and trucks, for everything operated by fossil fuel now. It will change all the service stations, and the snow that's plowed will remain clean and the water falling on your garden will not be acid. Corporate downsizing will be a feeble joke. The old behemoths may keep that up, but there'll be so many new jobs and corporations to do the conversion, you shouldn't notice.

And you may save the species *Homo sapiens sapiens* from an embarrassing extinction brought on by itself.

You know that I like quotations. Here's an archaism, something that isn't true anymore but can be again. It's from Thomas Hardy: "Who can say of a particular sea that it is old? Distilled by the sun, kneaded by the moon, the sea is renewed in a year, in a day, in an hour." Old Thomas could never have imagined that a century later, the Black Sea would be going anoxic and becoming devoid of life.

And here in closing is another one, from a remarkable contemporary named Dee W. Hock, retired founder of VISA, and now working in Rockland:

"Can there be institutions," he asks, "which have inherent in them the mechanisms for their own continual learning, adaptation, order, and evolution, and the capacity to co-evolve harmoniously with all other living things to the highest potential of each and all?" And he replies, as I echo him, "I simply do not know. At such times as these, it is no failure to fall short of realizing all that we might dream; the failure is to fall short of dreaming all that we might realize.

Mercator Miasma

Unitarian Universalist Sermon, 2 June 1991

GERHARD KREMER WAS ARRESTED on suspicion of heresy in 1544—he was accused of leaning toward Protestantism two years before Martin Luther's death—so he fled what then passed for the Low Countries, went to the border town of Duisburg into present-day Germany, and changed his name. If you look up *kremer* in a German–English dictionary, it is defined as a small shopkeeper; if you look it up in a German dictionary, you will find it defined as a materialist and a petty pinchpenny.

The name he chose for himself in Germany was much grander—*Gerardus Mercator*, thereby promoting himself from shopkeeper to merchant. In Duisburg he continued to work on his map of the world, based on a projection that had been in use for a couple of decades. In 1569 he published the map; it's the one still hanging in hundreds of thousands of schoolrooms to this day.

Mercator had the best of intentions: his objective was to help navigators; in fact, Mercator entitled his map *Carta ad usum navigatium accommodata,* Adapted for the Use of Pilots. Voyages of discovery were at their height during Mercator's youth. His projection, which turned a round earth into a flat map, made it possible for sailors to draw straight lines for their courses—the curved meridians that came to a point at the poles were laid out flat. Their intersections with the parallels were all at a convenient ninety degrees.

But. If you take a two-dimensional piece of paper and wrap it around a sphere, weird things can happen, and the weirdnesses occur according to where you have the cylinder touch the sphere. If you use Duisburg in Germany as the center of the world, the weirdnesses wax mightily in one spot and wane mightily in another. The easy one to spot is Greenland, which is bigger than South America on a Mercator map. In real life, Greenland's 840,000 square miles are less than an eighth of South America's 6.9 million square miles. If you scissor South America out of a Mercator map and lay it over Europe, the continent fits nicely within it. Yet South America is actually twice the size of Europe.

On we go. Mercator's Africa looks as though it could be tucked in a corner of the former Soviet Union. But Africa is 11.6 million square miles in size; the former Soviet Union is three-quarters of that area. It was those straight courses across the oceans that held Mercator's interest, not sailing around the Soviet Union or kayaking up the Nile.

As a result, for more than four hundred years schoolchildren and adults have been inculcated with Mercator's distortions of the world. We have the warped idea that north of the equator lie the most important parts of the world, little realizing that there is more than twice as much land in the south as in the north—18.9 million square miles in the north and 38.6 million square miles in the south. British schoolchildren see London as the center of the world—one goes west to Nome from England and east to Kamchatka. Similarly American schoolchildren go west from Kansas City to Mongolia and east to Tibet.

Let's go back a moment and set the world stage before the entrance of Mercator and his map. First, stimulus in the fifteenth century: Gutenberg (who changed his name from Gensfleisch, *people's meat*, and who can blame him?), credited with the printing press and movable type before his death in 1468, made multiple copies of documents and maps possible. Then came fifty years of explorers: Cao coasting along West Africa to the Congo in 1484, Diaz rounding the Cape of Good Hope in 1488, and da Gama making it to India by that route in 1498, Cristobal Colon (the Americans changed his name to Columbus because Colon didn't translate well) sailing to the

West Indies in 1492, Giovanni Caboto (John Cabot to you and me) to the Maritimes in 1497, Balboa on a peak staring at the Pacific in 1513, Cortez conquering Mexico in 1519–21. Rounding out the century, Pizarro conquering Peru in 1532.

There was Queen Isabella of Spain, who applied to the Pope to make the Inquisition a Spanish state enterprise in 1478; after seeing Columbus off, she completed driving all the Jews out of Spain. She got this idea from her husband's great-great-great-great-grandfather, Ferdinand III, known as "The Saint" and actually canonized in 1671. Back in the middle of the thirteenth century, he drove out all the Moors, suffered an economic collapse, and had to invite the Moors back so that somebody would do the work.

Three more items influenced Mercator: In 1529 the Muslims abandoned the siege of Vienna, and it looked as though Europe might be Christian after all. A Pole named Mikhailov Kopernik (who became Copernicus) came up with the bizarre notion that the earth revolves around the sun in 1543. Heeding the results of the Inquisition, Copernicus took two precautions: He dedicated his proposal to the Pope and delayed publication until he was on his deathbed (how wise he was to do so was proved sixty-seven years later, when Galileo corroborated Copernicus, got hauled before the Inquisition, and was forced to recant). Finally, in 1568, a year before Mercator published, the Netherlands revolted, threw off Spanish rule, and turned Protestant, further ensuring that Europe would be Christian.

You will recognize from all this that 1500 was a pivotal point. Prior to that century there had been five hundred years of the Dark Ages, when the European peoples' lives were nasty, brutish, and short. During that time the Muslim world was at its peak, busily inventing the zero, codifying the algebra hinted at by Hammurabi twenty-odd centuries before, mapping the stars, naming them and computing their magnitudes, and, incidentally, inventing the cap and gown.

In 1500, however, began the era of Europeanization of the world, five centuries of conquest of the other (and minor, by Mercator's projection) parts of the world. The United States joined in this enterprise as soon as it could, starting with the Monroe Doctrine. So now, as the year 2000 approaches to end a five-hundred-year period, we are ready for a new world order.

To achieve this new world order, we mustn't pay any attention to Mercator any longer and we must dispel the delusion he created. We must pay attention to the wonderful fact that the world is tipped 23.5 degrees in relation to the sun, whose true purpose in life was discovered by Copernicus. As a consequence, the 47 degrees of latitude between the Tropics of Cancer and Capricorn receive more than enough solar energy every day, year-round, to provide all the world's energy needs forever—well, for 4.5 billion years anyway, at which time the sun is scheduled to die out.

Today there are power grids interconnected all around the world. With the relatively high-temperature superconductivity that is on its way, plus other existing and improving technologies for power transmission, a world grid from natural sources within the Tropics becomes possible.

The mix within the Tropics is solar, wind, biomass, and Ocean Thermal Energy Conversion (OTEC). This last occurs because there are vast oceanic areas within the Tropics, already mapped, where the temperature differential between the bottom and surface water is sufficient to produce a substantial net gain of energy. The fact that the bottom water is nutrient-rich is also helpful; it's just the right stuff to make sea animals grow (Georges Bank is a natural OTEC; the Gulf Stream pushes masses of oxygen-rich water up over the bank, which is why the Gulf of Maine once produced 10 percent of the world fish catch). OTEC platforms, using infinitely renewable energy, can provide the power to operate fish farms, as well as the power for homes and towns.

Within the Tropics, the countries are not all that crowded.

Europe is approaching 270 people per square mile, not far behind Asia's 287, but Africa within the Tropics is only about 50 people per square mile, Central America and the Caribbean about the same, 44 per square mile in South America, and only 8 per square mile in Oceania, including Australia. Even the southern part of the former Soviet Union is around 30 per square mile (about like Maine in 1950). It becomes inescapable that the lands and waters between the Tropics are the ones that will save our ecological bacon. They can heal the world.

And the ambitions of the developing countries, most of them within the Tropics, should not be encouraged to imitate our excesses. They are not banana republics; they are our salvation. The knowledge perforce must flow from the developed to the underdeveloped, but it must be the new knowledge, not a repeat of our "successes," accomplishments like the Aswan Dam in Egypt, now viewed as an agricultural and fishery catastrophe, or the endlessly polluting freeways of Los Angeles, and so on down the dreary list. We have other technologies to export, like $1.19 reflector ovens (which lower the need for fuel wood) and solar-powered pumps and desalination plants (resulting in less drain on aquifers). Million-acre seaweed farms are already on the drawing board to furnish natural fertilizers ashore. And good, renewable, clean electric power can come from the Tropics via photovoltaics, from the trade winds, from biomass—soon you won't be seeing ads for the all-electric house, but for the all-electric country. And the electricity won't come from noxious coal plants or dangerous nuclear plants. Thus is fulfilled the Iroquois chief's admonition that every decision must be accountable for the health of the seventh generation.

Ah, but we shall merely be trading dependence on foreign oil for dependence on foreign electrical power, and indeed, dependence on a bunch of Hottentots and Hispanics, or so they will say in Washington, D.C. No. We know already that we need not be dependent on foreign oil, with just a few easy and sensible moves right now, from forty-mpg cars (Volvo already has one that gets seventy mpg) to the right light bulbs. With conservation and proper technology, the U.S. can be in balance with the rest of the world.

On my last trip to Nicaragua, one of my spectator sports was watching two remarkable young men who could build anything and, if the right tools were not to hand, build the jigs and forms and adapt the available tools to build anything. And they were teaching their confreres and the indigenes to do so, too. What have they been doing lately? At an inspiring combination farm and industry in Ohio, the two young men are building solar homes that need no backup generators by tucking the homes into the ground and powering them with photovoltaic heat pumps extracting heat from the heat reservoir in the ground. Someday they hope to use hydrocarbons only for lubrication in their maintenance shop.

The two Nicaraguan men make then-President George Bush's New Energy Strategy—which was not new, had little to do with conserving energy, and was a disaster as a strategy—literally look silly. To quote Fritz Schumacher, late of *Small is Beautiful*, "It is desirable to leave perplexity behind and get to work." The two young men are prime candidates to start the needed transfer of knowledge and technology and philosophy to that part of the world which Mercator helped us dismiss for more than four hundred years.

And in case you get in conversation with somebody from the New Right, point out that there's money in this shift, and a future that makes the bitter arguments about closing a military base seem less than trivial. There is work enough for everybody, right in and for the very areas really essential to a decent environment.

Fifty-five years ago, my college roommate and I were riding bicycles somewhere around Ulm and Augsburg on the way to Munich (we were a year ahead of Neville Chamberlain). To the right of the road, I saw a grassy wagon track going downhill.

Good lunch spot, I said; maybe there's a brook down there where we can sponge off. So down we coasted, eyes down, looking for the brook. At the bottom of the hill, we looked up to see two wooden towers and high fences and two machine guns in the towers, one gun tracking each of us. We had stumbled upon the back gate to Dachau. We fled.

In the next town we did the usual search for the German equivalent of a diner, trying to find one without the sign in the window, "Juden sind nicht hier erwunscht," not easy to find in 1937. While we ate, we described our experience to the proprietor and his wife and asked what was going on. "We don't ask," they said. "Something to do with the government. We don't know."

Well, this time around, with no boundaries to stop the acid rain, the ozone depletion, the chlorofluorocarbons, we should ask. We should look at the right map, because we *do* know what is going on.

Climate Change

PETER SHELLEY

"GLOBAL WARMING" IS ONE of those topics that challenges our very notions of what it is to be human and practice a sustaining human ecology. How can it be that my behavior—driving my car to the corner store for the Sunday New York Times or lighting of my house against the winter's gloom—could warm up the entire planet? How can it be that we as a species, in the name of progress, may have unleashed a chain of events that will alter the world for thousands of years to come? How is it possible that the price tag for our miraculous technological progress and quality of life advances may ultimately be significant structural decline of the planet's capacity to support our own species?

Edward—and it was always Edward to me—Myers wrote me many times during our fifteen-year friendship. It was a rare piece of correspondence that did not point to the threats of global climate change and humankind's pernicious denial and dissembling regarding those threats. Rightly proud of his miserly Honda hybrid automobile (70 miles to the gallon!) and forever shaking his fist, both symbolically and in real life, at the power plants that belched greenhouse gases into the atmosphere, Edward was not one to pull a blanket of feigned ignorance or helplessness over his head. He pointed the way to solutions. While he told me more than once "we're all doomed," the Edward Myers I knew had long before adopted a personal philosophy of living as if individuals actually could make a difference.

Current Scientific Consensus on Human-Induced Global Warming

There was a time in recent decades when there was some debate over the question of whether average global surface temperatures were increasing. That time has now passed: virtually all scientists now agree that average global surface temperatures are increasing and that change is accelerating over the historical record. Globally, the decade of the 1990s was "very likely" (90–99 percent probable) the hottest since 1861, when instrumental weather records first were kept, and was "likely" (60–90 percent probable) the hottest decade in the northern hemisphere in the last 1,000 years. Not just temperatures on the surface of the earth are rising: the lower atmosphere is warming. There is a 10 percent decrease in snow cover in the northern hemisphere since the 1950s; mountain glaciers are retreating; the period of ice-cover in northern lakes is shortening; spring and summer sea ice have declined 10–15 percent since the 1950s; there has been about a 40 percent decline in the thickness of summer and autumn Arctic sea-ice thickness; and global average sea levels rose somewhere between four and eight inches in the twentieth century.

Similarly, there was a time when there was a heated debate on the question of whether human activities or natural forces were contributing more to global warming. Again, most scientists now—-including many of those who once were skeptical—have come to believe that greenhouse gas emissions from human activities are principally responsible for these accelerating warming trends. Or as stated more precisely by the science panel established by the United Nations Environmental Programme and the World Meteorological Organization: "most of the observed warming over the last fifty years is likely [66–90 percent chance] to have been due to the increase in greenhouse gas emissions."

One can read and study these scientific reports and become quickly numbed or paralyzed: we are doomed. To some degree that is true: ecosystem changes have already been launched that will continue even if all further emissions were somehow eliminated. Forecasting models developed by climate scientists, for example, suggest that even with stabilization of these gases at present levels, the atmosphere will continue to warm at a rate of "a few tenths of a degree a century." Without stabilization, the models predict on average "several degrees per century." And even small changes matter. A long-term, sustained, localized warming of Greenland of only about 3° C would melt the ice sheet and raise ocean levels more than 21 feet. True, this would happen over millennia, but the face of our civilization would be unalterably changed.

In New England the impacts of global warming would be dramatic for both the region's two natural resource bases—our forests and our regional marine resources—as well as for coastal developments. Temperature changes in New England are estimated to increase from 3°–5°C by 2100 if current rates of greenhouse gas emissions continue, converting Boston temperatures to those between Richmond, Virginia, and Atlanta, Georgia. Such temperature shifts and the climate variability and increasing storm activity associated with the temperature changes will harm public health, leading to increased smog and acid precipitation. In addition, we can look forward to dramatic changes in composition of our regional forests as cold-weather trees move northward or die out due to insect predation and nutrient depletion. For Maine, the future offers dramatic changes in the Gulf of Maine ecosystem, long considered one of the world's most biologically productive marine areas. With increased warming there will be more frequent and severe toxic algal blooms, loss of critical wetlands, and clear changes in marine species abundance and type.

Greenhouse Gas Emissions Are a Significant Cause in Global Warming

While there are many constituent causes associated with global warming, a principal culprit is carbon dioxide (CO_2) emissions. In the words again of the Intergovernmental Panel: "emissions of CO_2 due to fossil fuel burning are virtually certain [greater than 99 percent probable] to be the dominant influence on the trends in atmospheric CO_2 concentration during the twentyfirst century." The current concentration of CO_2 in the atmosphere "has not been exceeded during the past 420,000 years and likely not during the past 20 million years." There are other human-produced greenhouse gases that are of critical concern, primarily because they are more effective than CO_2 in trapping warmth in the atmosphere: methane (CH_4)(twenty times more effective, ten-year trend declining), nitrous oxide (N_2O) (three hundred times more effective, ten-year trend increasing), and halocarbons such as CFCs and PFCs (most potent of all, variable trends—mostly declining).

Americans produced roughly 24 percent of the total global carbon dioxide emissions in 2000, approximately twice the emissions of the next highest global CO_2 emitter, China. Within the U.S. in 2000, 97 percent of the carbon dioxide emissions came from the combustion of fossil fuels: 37 percent of those emissions came from the process of electricity generation (coal, oil, and gas); 30 percent of those came from transportation-related sources such as automobiles, trucks, airplanes, and trains. Clearly, if we are to have any impact on reducing this critical greenhouse gas and reversing the climate warming trends so apparent today, the focus has to be on amending our transportation systems and our use and production of electricity.

Reversing Global Warming Trends

Still, we are hardly doomed. While warming will inevitably continue—-and ecological shifts and public health consequences may continue to escalate—-the trend can be reversed by reducing global and national emission rates. Because of the accumulation factor, however, the longer we wait to start, the more extreme the cuts will have to be if we want to avoid some of the more dire predicted futures.

The approaches to reduce carbon dioxide emissions (many

of which will also reduce other greenhouse gas emissions) are obvious: increase energy efficiency in all sectors, continue the shift to electricity production with fewer or no greenhouse gas emissions, increase cogeneration of heat and electricity in industry, and increase vehicle efficiency and improve fuel standards. In the longer term, solutions will come through reduced dependence on the automobile as a transportation source and shifts to cleaner and more efficient energy sources.

The New England Governors and Canadian Premiers have already made a strong move to jump on this sensible bandwagon, adopting a regional Climate Change Action Plan in August 2001. Under this plan, the governors and premiers committed to achieving a short-term goal of reducing regional greenhouse gas emissions to 1990 emission levels by 2010; a mid-term goal of reducing regional greenhouse gas emissions to at least 10 percent below 1990 levels by 2020; and to achieve a long-term goal of reducing regional greenhouse gas emissions below the level that poses a threat to climate, which current science suggests is in the range of a 75 to 85 percent reduction below 2001 levels. The plan further specifies nine "action steps" to which the states and provinces are committed.

One analysis indicates that implementing an aggressive program of global warming solutions over the next decade could reduce primary energy use by 19 percent and carbon emissions by 31 percent below that which would occur under a status quo approach. Renewable energy production could increase by 48 percent, with wind and biomass generation providing about 15 percent of total primary energy generation. Industrial cogeneration of electricity could reach 26 percent of projected energy demands in a decade. If these changes took place, they would bring the New England region 16 percent below its 1990 level of emissions, well beyond the 7 percent reduction target set by the Kyoto Protocol, adopted by the Parties to the United Nations Framework Convention on Climate Change in December 1997 (rejected by President Bush in 2001).

Total cost to New England? Annual energy bill reductions are estimated to approach $305 per person in ten years with cumulative net savings to the region of $31 billion in ten years (1998$). Employment associated with implementation of the program is estimated to be 41,200 new jobs.

I think that Edward Myers would be pleased to know that New England is already contributing less per capita to global warming than the national average; it would appeal to his Yankee parsimony. While it's hard to imagine him giving much credit to regional state and provincial governments leading the charge on addressing climate change, he would no doubt be impressed that they had even stuck their political necks out on this one. I can well imagine a flurry of letters off to the New England congressional delegation in Washington calling their attention to the national disgrace of the present United States policy on global warming and attempting to shame Congress into at least matching the State of Maine's initiative, if not raising the ante. As Edward lamented in a recent global warming-related letter to the Conservation Law Foundation on April 16, 2002: "There's a lot to do, and I wish I were sixty instead of eighty-five." So do I.

PETER SHELLEY is vice-president and former director of the Maine Advocacy Center director of the Conservation Law Foundation. In 1955 Mr. Shelley was awarded a Pew Charitable Trust fellowship for his marine advocacy work in the Gulf of Maine. Subsequently, he helped establish the Northwest Atlantic Marine Alliance, based in Kennebunk, Maine.

TWO

NATURE

The Festival of Lights

First appeared in Working Waterfront, *April 1999*

NEXT TIME AROUND, instead of picking a hamlet wherein the house and dock building are smack dab in the middle of the cove's shore and facing the southwest quadrant, we'll see if we can't at least dicker for one of the points, because as it is now we are surrounded by lights which take away the night sky. We elderly types grew up with the whole sky ready and to hand (in fact, for heaven's sake, we can clearly recall as a child a man on a bicycle who came around every dusk to torch off tiny gas lights).

There's a fellow up to York Harbor who feels the same way, pointing out that Yale, in New Haven, Connecticut, had to move its astro-observatory to Chile to get away from the lights of New York, New Haven, and Hartford. It's the same here. We used our sextant to discover that the sodium and neon glare of Boothbay Harbor, 4.5 sea miles away across the river and to the southwest, wipes out 20 degrees of the night sky in summer and only a few degrees less in winter. Brunswick Naval Air Station, Bath Iron Works, and Maine Yankee combine to take away 15 degrees of the night sky, at least. A couple of our children never knew that there were any stars from S by W to W by N on the compass until they sailed the Atlantic Ocean to Ireland several years ago.

In an earlier column we alluded to the helpful type downriver who saw our running lights one night as we were grabbing the last of the flooding tide to get through the Narrows and quickly produced a galaxy of floodlights on his dock. The lights destroyed our night vision, which had formerly picking up every lobster pot and drift log. The dark was light enough.

Now we get to our neighbors, wonderful people all of

them, but mostly of a generation susceptible to the marketing blandishments of Central Maine Power.

First is a house of a gallant lady wintering alone while her husband is in the hospital. Mounted on the northeast corner of the second floor is a powerful spotlight aimed accurately at our bedroom and guest room. This light usually is turned on at dusk to greet a caretaker, who looks in daily, and to speed his departure. The problem is this: the beam is 135 degrees from the woman's back door and only a pale edge hits her driveway, while the light fully illuminates about two acres of our hayfield, along with the front of our house. This light was right in line with the transit of Venus, which fascinated us some weeks ago, but we got around the interference by using some scraps of roofing paper with small flaps cut in them to view the sky, though that did take the joy out of the edge of darkness and a rare astronomic event.

Our neighbors to the east have two lights to limn a highly active parking lot for their family, and they need them. Their modest orbs do a number on our view of the rise of that fine old Irish constellation O'Rion. Happily, the neighbors acquired a bigger sloop last summer, and it winters between us and the lights from the end of October to May, and that solves that.

To the north, for the last twenty-plus years, alas, is a streetlight. This came about because a Christmas wreath was stolen in a drive-by incident. You don't blame anyone for being upset; being robbed, no matter what the value, leaves you unsteady for a long time. Never mind that the town's paid $2,425 and counting to CMP for the streetlight.

We have a small skating pond and vernal pool right in line with Polaris and with the streetlight. We and the children used to set the record player (you remember, those gadgets before CDs) at the living room window, put on the Brandenburg Concertos at low volume, and skate under the stars. No more. Then after the ice went out (March 5 this year, the earliest in history), we'd explore the rites of spring at night—frogs and peepers, a few turtles, pollywogs, and a zillion baby leeches. You can't get near them at night anymore, because the streetlight gives away your slightest move and the wildlife goes quiet. It was decades ago, but we still remember a quiet night when a couple of Lesser Yellowlegs whistled in and we could follow their every move by the reflecting starlight.

The streetlight's on a ninety-degree corner and its beam masks the oncoming headlights we used to be able to see as we made the turn at night coming home. Occasionally a couple of young cowboys in the neighborhood take the curve a bit wide; you have to be alert now.

We wish there were a way that those people who have given up the loveliness of the night could somehow feel safe and still without the lights, as in Gray's "Elegy"—"leave the world to darkness and to me."

Reprinted with permission of The Island Institute, Rockland, Maine.

“What Are Your Instructions?”

Baptist Church Sermon, August 1984

YOUR PASTOR, AS USUAL, first enlisted me to lead today’s worship service and then, also as usual, warned me to stay away from politics. Previously, I’ve observed that politics came from the Greek word *polis*, meaning town or community, and surely there was no danger in that. This time, I have at the ready a quotation from Henry David Thoreau, a curmudgeon after my own heart:

> To one who habitually endeavors to contemplate the true state of things, the political state can hardly be said to have any existence whatever. It is unreal, incredible, and insignificant; to endeavor to extract the truth from such lean material is like making sugar from old rags. Generally speaking, the political news, whether domestic or foreign, might be written today for the next ten years with sufficient accuracy.... “But will the government ever be so well administered,” inquired one, “that we private citizens shall hear nothing about it?” The President answered, “At all events I require a prudent and able man who is capable of managing the affairs of state.” An ex-cabinet minister replied, “The criterion, Sir, of a wise and competent man is that he will not meddle with such trivial matters.”

So of course you are on safe ground with me.

The scripture readings this morning are all from the Lectionary for this Sunday. First the psalm of praise we read together, a recognition of an omnipotent God ruling land, sea, and air and a prayer to him for strength and peace; then two seagoing tales, both familiar, one of Jonah and one of Peter, each loaded with humanity.

Jonah was running away from his instructions, as you will recall; God had asked him to go to the evil empire of Ninevah and tell them of God's plan to destroy it, and Jonah thought to get out of the job by taking ship for Spain; after being returned by the whale, Jonah repents, agrees to go to Nineveh and tell them the bad news. And then he is perfectly furious with God for changing his mind and sparing the city. Nothing makes people more put out than being prophets of doom and then having things turn out all right for their enemies.

The episode of Peter is also loaded with human qualities, very recognizable. Here is the man—Peter, Petra, Pierre, the Rock—selected by Jesus to be head of the church, who a few weeks hence will deny him thrice before cockcrow, who asks to walk on the water, does so, then does a classic double take that is almost comic—"What on earth am I doing walking on the water?"—begins to sink as his faith departs, and calls for help. He had his instructions and forgot them, but the Lord buoyed him up, chided him for his failings, and did not change his mind about his earthly successor. Jesus did not expect perfection. Only faith. And Peter went on to be a wide-ranging influence on the temporal world, which is none the less God's, and followed his instructions.

In a much later day, we can listen to a Wampanoag Indian lady speaking about the conflict between the gentleness of her tribe on Cape Cod and what that area was fast becoming. This is what she said:

> My people do not understand you or why you do the things you do. Why are you still trying to take more of our land? Why must you own things? Why must you always have more? A seed, a flower, a tree unfolds according to the instructions it has been given. We have always tried to live by ours. We don't understand yours. How have you been taught to live? What are *your* instructions?

How do we reply? With my fondness for Greek words, I begin with *oikos*, meaning household or home. It is the root word of economy, about which we hear and, alas, think so much these days, which originally meant the management of a household, *oikos*. It is also the root of ecology, about which we should hear and think more. When you use the word *environment*, it includes everything but us; it is our objective surroundings, something we're inclined to think we can manage. When we use the word *ecology*, meaning the world household, we know that we are included, and the very inclusiveness implies interconnectedness. Human interaction isn't with the natural world, but is within it and part of it. If we seek the truth, it is a truth held in tension between paradoxes—our having been chosen as the superior species under God, on the one hand (or so we may think), and being subject to all the processes of the planet earth, on the other.

On the last day of July, Whitney Carter of The Carpenter's Boat Shop and I decided to move the hulk of a forty-three-foot sardine carrier to the cove of an island a couple of miles away. We thought it best to do it at night, for a better chance of calm water. Why is it calm inshore at night? Because when the heat of the sun has gone, the temperature of the land and of the sea are nearly equal, and the water can therefore be smooth as glass, which it was. Why that night? Because the moon set at 10:30 P.M., only about two hours after sunset, and the combined power of the sun and the moon located on the same side of the earth gave us an eleven-foot-plus tide, almost two feet above normal (the sun's tidal effect is about one-sixth of the moon's but it helps) and that would let us beach the vessel where it would never float again (if we remembered to pull the bilge plug, and we did). We used a 250-pound aluminum skiff with an eighteen-horse outboard as the tow boat for the nine-ton hulk. I won't go into the miracle scientists call physics to explain how that is possible—nay, easy.

The vessel floated off our shore at 1:00 A.M.; we had

already picked our stars for the course. Aldebaran or some such was just right for south, which is where we were going, so of course Polaris was right in place as a stern-bearing for the tow. Why the North Star, millions and millions of miles away, should have such a tiny orbit as to be dependably north at all times is another miracle. As it turned out, the dark was light enough, and we could avoid the lobster buoys and the drift logs that the big tides bring in without destroying our night vision with flashlights.

Well, we got there a half-hour before the top of the tide, warped the vessel in, snugged her up with a come-along fastened to an oak tree, and kept her upright with two breastlines to other trees. The fact of a boat's balance point is another miracle: the lines have a breaking strength of only 3,000 pounds, but they're sufficient to keep nine tons put if she's balanced properly on the keel.

That night the cove was full of the flashes of small herring schooling over their new territory made by the spring tide. At 3:30 A.M. we took a break and sat on deck for a thermos of coffee, congratulating ourselves on the work we had done, as though we had done it, and looked up to see the sky full of Northern Lights, green and red and yellow flickering all over the place—another unexplained mystery. "Too much," was all we could say.

We loaded the pumps and the tools and the spare lines in the outboard for a quick trip back to our dock, and then headed right back to the island for the first light of dawn, so that we could see how the vessel was settling in the falling tide and prop her up with timbers. The last trip back, with flotation logs and deck gear, was in full daylight at 6:00 A.M. I tell you all this because for eight hours, although we did have to use three gallons of fossil fuel and did have to converse a little bit, we were in otherwise total silence and totally interconnected with the major physical forces of the Creation. It was a glorious night's work, and rewarding, far beyond the little transfer we accomplished. I recommend it to you, perhaps not in the same detail, but in the same attention to the mysteries and the comforts.

We must no longer look at God's ecology from Johannes Kepler's view, stated in 1605 that "the celestial machine is to be likened not to a divine organism but rather to clockwork." His was the first step along a path through Descartes to Isaac Newton's mechanistic world-view that has us treating all matter, including the living world, as lifeless and apart from mind. Possibly that path was necessary in order to reach the turning now before us.

There are signs of this change. Dr. John Wheeler, summer resident of South Bristol, dealing with quanta and black holes and other astronomical puzzles, concludes that, "In some strange sense the universe is a participatory universe." Darwin probed a little earlier, saying, "The grand question...when dissecting a whale or classifying a mite or a fungus is 'What are the Laws of Life?'" Gregory Bateson wondered, "What pattern connects the crab to the lobster and the orchid and the primrose and all the four of them to me? And me to you? And all the six of us to the amoeba in one direction and to the genius in another? What is the pattern that connects all living creatures?" And Chief Black Elk said, "All life is holy and good to tell, if only we choose to listen." Yet economic and political strategies continue to be based on assumptions of indefinite exploitation and continued growth. At that level there is no talk of healing. To focus on healing, we can embrace the interconnectedness of all life. We can heal the tear between the sacred and secular, and again view the totality as sacred.

Jonah was healed of his anger at not having his way in destruction; Peter was healed of many failings both on the Sea of Galilee and in the Garden of Gethsemane; the psalmist asked for the healing of courage and strength. Jesus' life was healing. Do we ask too much of ourselves if we try to take on the same task, of healing the sacred earth? We have our instructions. We have had them for a long time.

E. F. Schumacher Revisited

Unitarian Universalist Sermon, 3 October 1999

THIS IS A TEST. In 1784 Thomas Jefferson wrote: "Indeed I tremble for my country when I reflect that God is just." Now, who said this? "I see in the future a crisis approaching that unnerves me and causes me to tremble for my country. As a result of a war, corporations have been enthroned and an era of corruption in high places will follow, and the money power of the country will endeavor to prolong its reign by working on the prejudices of the people until all wealth is aggregated in a few hands and the Republic is destroyed." Answer: Abraham Lincoln, 1864.

In 1886 Chief Justice Morrison Waite wrote the majority opinion that corporations were to be considered as persons under the Fourteenth Amendment's equal protection clause. That was the year of the Haymarket Riots in Chicago, which led to the formation of the American Federation of Labor and the United Mine Workers. The Supreme Court didn't pay much attention to these events, and in 1919 further solidified the notion of corporations as endowed with eternal life and immunity for their behavior.

And now, to population. The growth of population began in the Fertile Crescent between the Tigris and the Euphrates Rivers ten thousand years ago. Turning to farming rather than hunting and gathering, this ancient civilization helped start the growth in the world's population that will reach, in 2025, approximately 8.8 billion people. The United States will have gone past 300 million by 2025 but it will have faded to less than 3.5 percent of world population and, unless there are drastic changes, will still be using 35 percent of the earth's resources.

E. F. Schumacher, known as Fritz, and a remarkable fellow,

expressed his fervent dissent with the modern situation until his untimely death on a Swiss train in 1977 at the age of sixty-six. His most famous book is *Small is Beautiful* but he should also be remembered for *A Guide to the Perplexed*, *Small is Possible*, and *Good Work*. A compendium of his thoughts fills *This I Believe*, published in the year of his demise.

Schumacher was born in Bonn, Germany, in 1910 and was a prodigy through school and university. He was appointed assistant professor of banking at Columbia University at the tender age of twenty-three. He worked in Wall Street and in the City of London, and returned to Germany to run a barter-based import-export company of his own devising; the company thrived even in the Great Depression. At the time Hitler had been in power for four years—in 1937—he picked up and went back to England. He was convinced that Germany had no future under Hitler, and that the rebuilding of Europe would be under British leadership; he hoped that he could take part in the needed economic and social reforms.

Instead, Schumacher suffered the hostility accorded all German nationals in England, was interned and sent off to Northhamptonshire as a farm laborer for a while, and then sent by the Crown to do research at the Oxford Institute of Statistics. When the war ended, he was sent back to Germany as economic adviser to the Allied Control Commission, and there was looked on as a rat who had deserted the Fatherland.

These details explain his description of himself as an outsider, who had been beset with poverty, social injustice, and alienation. Because so many viewed him as an impossible dissenter with a Germanic name, he used pseudonyms when asked to write for the *London Times* or the *Economist*. He more or less designed and laid out the ideas for the Marshall Plan and the Beveridge Report on Full Employment, but never got any credit.

As a commuter for twenty years from his suburban home to London, he spent the train time studying comparative religions, especially those from the East. Over the years, he progressed from a youthful Marxism to Buddhism and from there to Gandhi, whom he admired for his search for truth and looking at matters whole. Finally, to the astonishment of many of his colleagues and readers, he turned to Roman Catholicism a half-dozen years before his death.

One of his epiphanies was making a distinction between poverty and misery. After all, a billion and a half people, a quarter of our present 6 billion, live on a cash income of a dollar a day. Poverty, relative or not, may be endemic, but misery does not need to be.

This led him to his rule of the disappearing middle. As an example, think of the World Bank or the International Monetary Fund financing a 600-foot-high dam which drowns 100 villages, produces a lot of electrical power for the cities, which helps the rich get richer, whilst the displaced residents of the 100 villages sink deeper into poverty. To work against the disappearing middle, Schumacher and his colleague George McRobie formed the Intermediate Technology Development Group, which now has headquarters in six countries. Instead of capital-intensive work projects, ITDG provides the aid of tools plus skills and education to rural communities in developing countries.

When George McRobie visited our mussel farm some years ago, he caught me in oilskins with a bandana over my mouth, lying on the dock under a thirties-era Maytag washing machine we had brought back from the dump and rebuilt into a mussel separator and washer. I was trying to clear a deep hemorrhagic infarct of mussel shell, mud, kelp holdfasts, and whatever else in the drain. I was ashamed, but George was absolutely delighted to see very intermediate technology under maintenance. Maybe today, somewhere in Kuala Lumpur, ancient washers are cleaning shellfish, and I have made my contribution to ITDG.

To Schumacher, "the art of living is always to make a good

thing out of a bad thing. This then leads to seeing the world in a new light."

And in that new light, he said:

> The problem posed by environmental deterioration is not primarily a technical problem; if it were, it would not have arisen in its acutest form in the technologically most advanced societies. It does not stem from scientific or technical incompetence, or from insufficient scientific information, or from a lack of information, or from any shortage of trained manpower, or lack of money for research. It stems from the lifestyle of the modern world, which in turn arises from its most basic beliefs.... It is only a complete metanoia in all departments of life, rather than engaging in an overextending battle with mere symptoms, which will educate us to change, to avoid universal breakdown.

My goodness. I'm almost done and I haven't really drawn in the fisheries. So here's another perfect example from Schumacher:

> Just as industrial society is fundamentally unstable, so within it the conditions which offer individual freedom are unstable. They are unable to avoid rigid organization and totalitarian control.... It is difficult to see how the achievement of stability and the maintenance of individual liberty can be made compatible.

That is exactly what has happened in our fisheries.

Schumacher was a wonderful and perceptive individual. A quarter-century and a bit more have passed since *Small is Beautiful*. But I recommend you find a copy and enjoy.

As you know, I like to conclude with quotations. Both of

these are quoted by Fritz Schumacher, both of them are from Aristotle.

The first: “To the size of states there is a limit, as there is to other things, plants, animals, implements; for none of these retain their natural power when they are too large...but they either wholly lose their nature or are spoiled.”

And the second: “The slenderest knowledge that may be obtained of the highest things is more desirable than the most certain knowledge of lesser things.”

Ed Myers: Mentor, Investor, Friend

Tom Chappell

THE TELEPHONE RANG next to my bed at 8:45 on a Thursday evening in 1971, our second year of Tom's of Maine. The call was from a man claiming to be the director of the Darling Research Center in Walpole, which was to open in two days, so long as there was drinking water.

Ed Myers introduced himself quickly and then dropped the news that the new artesian well just drilled for the Center was already contaminated with gasoline from a nearby underground storage tank, left from the original days when the property was a dairy farm. Scientists from all over the world were to arrive at the site in two days for the opening of the University of Maine's new marine research center.

Ed's manner was serious and professional, suitable to the crisis situation. He commented in a wry tone that he was "concerned that the scientists might not appreciate tap water at the lab benches reeking of gasoline." I was intrigued by the chuckle in his voice.

Ed informed me that he knew I was the only person in Maine who could possibly come up with a solution for eliminating gasoline from soils and well water. Here I was, not sure where the next payroll for my company would come from, and this man was portraying me as a cleanup expert. I suspected that, in his creative imagination, he had zeroed in on the company's new non-phosphate laundry detergent, Clearlake, as his salvation. Clearlake, the first nonpolluting detergent that cleaned clothes and kept waters free of harmful phosphates, was just reaching health food stores around the country. I was happy with it as a detergent for clothes but had never considered it in connection with underground gasoline leaks.

"Could you come to Walpole tomorrow to look at the

problem?" asked Ed. Still keeping my distance I responded that I had other plans. Urging me to reconsider, he said every day was critical and he must have a solution before the scientists arrived over the weekend. I pondered: Ed sounded intelligent over the phone. He displayed a sense of humor. He needed help and he was open to new solutions.

"I'll be there early in the morning," I said. We agreed to meet at the Damariscotta Diner for an early breakfast and briefing. Before signing off, Ed asked, "What is your fee?" Without much thought I said confidently, "A thousand a day plus the Clearlake." After a moment of silence, Ed replied, "May I ask about your credentials.... And your education?" "Sure," I said. "I am an entrepreneur with an undergraduate degree in English from Trinity College." More silence. "Just right," he exclaimed. "See you at 6:00 A.M."

What a set-up! A Princeton graduate who had succeeded in his own business, Saltwater Farms, which shipped live lobsters and seafood chowders around the country, and a budding entrepreneur from Trinity College looking for anything that might keep his new business going. Two peas in a pod, one grayer than the other, acting as if they knew what they were doing! Role playing! Having a ball!

After a good breakfast the next morning, our minds were attuned to the problem. We visited the site at the Darling Center. "Could we dig," I conjectured, "concentric circles in the ground around the gasoline storage tank with a backhoe, mix a 55-gallon drum of Clearlake into a tank wagon of water, and spray the liquid with a fire hose in the concentric circles? We would need to pump the water like hell simultaneously out of the well into an empty tank wagon to draw the detergent through the soil and out through the water in the well." Ed looked distinctly worried.

"Well," he said, "that seems like a plausible solution." Both of us knew, without admitting it aloud, that the idea was crazy. We were just going to wash the earth clean of gasoline. Nothing to it! We both smiled at the thought of success, because it's crazy ideas and the vision of what's possible that make entrepreneurs tick.

That night I wondered if Ed would scrap the plan after giving it a little more Princetonian logic. But he was a creative man and an entrepreneur at heart. "All is in place and ready for the Clearlake," he reported to me the next morning on the site. Tank wagons and backhoes were ready to go.

The washing began at 9:00 A.M. that morning. At the outset, the gasoline contamination was at 200 parts per million. Turn on a faucet and the strong fumes were enough to cause rapid dizziness. After three hours of hosing down the ground with Clearlake and water, and pumping the artesian well steadily, gasoline levels had dropped to 25 ppm. Not enough yet to open the research doors, but great progress! By late afternoon, the count was down to 5 ppm and the gasoline smell was indiscernible. It had worked! Ed could open the facility the next day.

As I was saying good-bye to Ed from my car that afternoon, he leaned over and asked with a smile on his face, "Is there any stock for sale in your company?" Huh? Stock, in my company? "Well, not at the moment," I replied, "but I'll let you know if that changes."

On my way out of Damariscotta, I stopped at a pay phone and called a friend, an attorney in Portland. Wonder of wonders, he was in his office at 5:00 P.M. on a Friday afternoon. I asked him how I could go about selling stock in my company. After a moment he said, "You'd better come in on Monday for a little conference." I knew I needed working capital to build my business. And here was a fellow I hadn't known at the beginning of the week, in a captain's wool cap, bow tie, and rubber boots, with whom I'd just pulled off an innovative rescue, who wanted to invest in my company, if the numbers were right.

I called Ed a few weeks later to say I would like to discuss his investment in Tom's of Maine. I spoke of our retail success with Clearlake and our possible new products of personal care products, such as soap and shampoo. Ed reported that the Darling Center was still free of gasoline contamination. And, yes, he would like to learn more about the company. We made a date for another breakfast at the diner.

"Your business plan sounds interesting, but how will I make a return on my investment?" asked Ed. Looking over my neat report, I saw that my prospectus had all the proper legal disclaimers, all the concepts for future sales growth, but no financial rationale to induce an investor.

"Well, what kind of return would motivate you?" I asked. Ed explained that venture investors typically like to achieve a 5 percent to 7 percent return during a seven-year period. "And how would I display that kind of return?" I asked innocently.

Ed looked me in the eye for a long moment, then turned over his paper placemat to scribble the format of a pro forma financial statement. One placemat turned to two, then to three before my teacher had laid out for me a profit and loss ledger, balance sheet, and cash flow statement. He showed me in a few short minutes how I could induce an investor to give me money in return for ownership in my company, thereby possibly creating wealth for both of us.

I looked at the overturned placemats for a few moments. "Ed, would you like to invest?" I asked. "I would," he said decidedly and wrote out a check. My new friend was now mentor, advisor, and partner.

I believe that people like me find people like Ed because in taking a risk, you are more open and vulnerable to seek help yourself, and the universe responds. I solved his problem and he solved mine. Through grace we found each other. I think Ed saw himself in me and re-enacted what he had learned as an entrepreneur by teaching me what I needed to learn. He gave me confidence, guidance, and money, all with a sense of good nature and humor. Ed didn't care about the investment of money as much as he did the joy of helping and coaching me.

Someone must have helped him, too. He remembered the law of the universe when he met me: pass it on.

Time passed and Ed and I stayed in touch. I was always grateful for his distinctive letters. In 2001, thirty years after our first meeting in Damariscotta, Ed came at my invitation to a small gathering of leaders interested in how values impact business character and performance. I had recently co-founded the Saltwater Institute, a non-profit education foundation dedicated to creating values-centered leaders and cultures. Tom's of Maine, still growing and prosperous, had been a learning ground for the intentional application of values to business strategies. I had established the Institute to help others learn how to put values first for enhanced business performance.

"What is it that brings you at this moment to be interested in values?" I asked the twenty-some business leaders at the gathering. Each shared a concern for the lack of values in our society and workplaces. When it came time for Ed to speak, he said he wanted to drive down from Damariscotta to Portland to see what his friend Tom was up to again. He wanted everyone to hear his warning that we don't have much time left to save our environment. Ed said emphatically that we must act now to decrease the level of emissions from petroleum distillates into our environment or there will be no reversing the world's degradation. Ed had left his cherished cove in Walpole for a few hours to support his friend, see what he was up to, and deliver his message.

Ed had appeared when I was taking another risk. His intersection in my life at that point shows that the universe provides when we follow our destiny. Two people meet, and the world is new and all is possible.

TOM CHAPPELL is the president and co-founder of Tom's of Maine, the author of *Managing Upside Down: Seven Intentions for Values-Centered Leadership,* and founder of the nonprofit Saltwater Institute.

THREE

OCEAN

Nations and Fish: Conflict Resolution on Georges Bank

Unitarian Universalist Sermon, March 1985

If the ocean were just a static body of water, a relief chart of the continental margin of eastern North America would be merely an interesting picture. But of course it isn't. The Gulf Stream scours its way around the end of Florida, and pours in a mighty river northeastward, diverted by the escarpment off Cape Cod, then makes its way in a curve to allow palm trees to grow on the southern coast of Ireland. The Labrador Current comes down around Cape Sable to enter the Gulf of Maine and the Bay of Fundy. Entering the Bay of Fundy, it makes a giant counterclockwise gyre, along the Maine Coast down to Cape Ann and Provincetown and onto the shallows of Georges Bank, whence it is urged back to Cape Sable by the edge of the Gulf Stream and around it goes again, in the springtime picking up thousands of cubic miles of runoff from all the rivers, thus changing the gyre's speed and temperature according to the season.

The reason that the Gulf of Maine is such a great place for fish is that the moving ocean comes up against the Cape Cod escarpment and there is a great upwelling of plankton-rich water pushed onto all the banks: Georges, Brown's, Sable Island, Banquereau, St. Pierre, and the Grand Banks, with the Flemish Cap as a sort of bonus. But the greatest of these banks is Georges, being nearest the Gulf Stream and having the cold waters of the Gulf of Maine behind it, and that has made it the most productive fishing area in the world.

Everybody who has fished here wants it all. Allegedly the Vikings came over in about A.D. 1000, running down the 45th parallel by the simple expedient of cocking their fingers at the North Pole and keeping it roughly in its place. They were set south by the Labrador current so their landfalls are suspected everywhere from Mt. Desert Island to Newport, Rhode Island. They were low on supplies by the time they reached this continent and stopped to handline cod.

Leaping ahead 500 years, we find a French Admiral Chabot ranging the area from 1527 on and collecting tribute from every vessel he stopped. It was not until 1632, 105 years after Chabot began taking admission charges for the high seas, that the British and the French in the Treaty of Saint Germain agreed that it was illegal.

Leap another century ahead and the British and French negotiated the Treaty of Aix-la-Chapelle. This was a bitter pill for the fishermen from the Massachusetts Bay Colony who had so fiercely and cleverly besieged the French fort of Louisbourg on Cape Breton Island. When the rabble in arms returned home to resume fishing and to enlarge their range of fishing eastward, they found themselves set upon by both the British and the French, since France had gotten Cape Breton back under the treaty.

In the interest of brevity, let's pass over the American Revolution for the moment and go to 1819, the first year of a subsidy on U.S. fish in order to keep Canadians out of the American market. It didn't work, like most subsidies, and it was followed by tariffs on Canadian fish, which reached 20 percent in 1852. That same year, Boston dealers complained that the tariff was too low: "The Canadians have expelled us from Catholic Europe and are now taking our domestic market."

Back in 1841 Daniel Webster ran out of money, as he so often did, and went on retainer to the British banking family of Baring to lobby for their causes in the United States. Things might have been okay if old Daniel and English Lord Ashburton (whose maiden name, so to speak, was Baring) had not been designated to settle the boundary between Maine and New Brunswick. Quite a lot of money exchanged hands, and out came the Webster–Ashburton Treaty, as a consequence of which Maine lost about 3 million acres to England (but that's another story!). Due to a long and merry negotiating evening, the U.S. also lost Deer, Campobello, and Grand Manan Islands to the other side of a boundary line drawn by Ashburton, which trailed off at Machias Seal Island, a lack of definitude that has risen to smite us to this day (there is still no official boundary between the U.S. and Canada for twelve nautical miles east of Machias Seal Island).

As an aside, just to prove the rule that it takes a century to resolve a fisheries dispute, Elihu Root, as U.S. Secretary of State, began drawing up his brief in 1908 for presentation to the World Court in 1910 to resolve some fishing anomalies left over from the Treaty of Paris in 1783 and not settled by the Treaty of Ghent in 1814. I use the word "brief" advisedly, as Mr. Root's presentation to the court had a 40-page introduction, a 100-page appendix to the foreword, and an argument of 433 pages. Nobody really knows who won, which is a standard condition of any fisheries negotiation: if both sides are angry, then it was successful.

Now a leap to the 1960s, when Georges Bank inadvertently became part of a Soviet Five-Year Plan. The USSR had a surplus of oil and decided that diesel fuel was a good trade for protein. Over came the Soviet pulse-fishers and the factory ships, dwarfing the elderly and relatively tiny draggers from the U.S. and Canada who were working on Georges Bank. The Poles and the East Germans and the Spaniards and just about everybody else followed the Russians, including the Japanese. The annual U.S. catch of haddock, rarely below 100 million pounds previously and sometimes as high as 180 million, fell to 19 million pounds. The offshore lobster fishery, just a few years old, found its traps swept away by giant foreign trawls. The Canadians working on Georges Bank were faring just as badly as the U.S. fishermen.

The U.S. Congress acted to protect the country's fish, passing the Magnuson Fisheries Conservation and Management Act in 1977. The Act claimed the water and submerged lands out to 200 miles from shore as a fisheries management zone, from which the U.S. could exclude foreign fishing vessels. The 200-mile limit was also put into effect by many other countries around the world. We all saw the "200-Mile Limit" bumper stickers on many a pickup truck, and we thought its inception was going to bring a new era of prosperity and privacy to American fishing. The shock came with stock assessments by fisheries scientists: by the late 1970s the population levels were so low that fishing quotas had to be established until the stocks recovered. Fishermen learned that the government which can do something *for* you reserves the right to do something *to* you. The golden age everyone expected became just the gilt edge on the mortgage notes for bigger vessels.

Then enter the Canadians. Their 200-mile-limit line started from Machias Seal Island and went in a straight line right down onto Nantucket Shoals, moving a vast area of the Gulf of Maine into Canadian control. The U.S. boundary line, invoking the equidistant rule used in the World Court in 1910, took various forms, even unto taking a slice off Brown's Bank, but ended up realistically (a State Department word) going out through the Northeast Channel.

Then the fun began. In negotiations, the Canadians argued that Cape Cod was a geological aberration that would disappear in 5,000 years and that therefore their boundary line should move 17 miles nearer the U.S. coast. The Americans explained that sure, there was a Canadian lighthouse on Machias Seal Island, but we had lent them the island as a kindly gesture so that their vessels could get up Grand Manan Channel without mishap. And pushed the boundary line back again, closer to Canada.

A lobsterman from Cutler was arrested by Canada for setting traps too close to Machias Seal; a draggerman from Deer Isle had his vessel confiscated for being a mile too close to St. Mary's Bay. Gear clashes abounded as both countries sent fishermen onto Georges Bank to prove they could. Canada averred that it would be the fifth-largest fishing nation in 1985.

In the course of all this, OPEC raised its burnoosed head, and the possibilities of undiscovered offshore oil on Georges Bank raised the stakes higher. There had to be a treaty.

So that's what Jimmy Carter set out to do. The man he picked for the task was one of those Washington super-lawyers who didn't know a haddock from a hydrocarbon, who was Washington counsel for the Canadian potash industry, and, worst of all, went to Yale.

The Canadians have long memories. The Tories who were exiled from the United States at the end of the Revolution and shipped off to nine months of winter and three months of cold weather in Nova Scotia were, in a sense, the best and the brightest. After the Revolutionary War there were suddenly more Harvard graduates in the uninsulated log cabins of Shelburne, Nova Scotia, than anywhere else in the world, including Boston. The Canadians seem to have kept this fact in mind during the negotiations two centuries later. What came out of them was an agreement so abrasive to U.S. fishermen that the American fishery ganged up on the Senate and managed to kill treaty ratification. Which brought all progress to a standstill. In the meantime, the basically friendly relations among the fishermen of both nations fishing in the Gulf of Maine worked out a grudging *modus operandi* on all of Georges Bank.

At the diplomatic level, the two countries agreed to submit the boundary dispute to the World Court in the Hague for binding arbitration. Reversing the wisdom of Solomon, the World Court cut the baby in half and drew a line which stops twelve miles away from Machias Seal Island. The court knew full well where it feared to tread. The boundary, known now as the Hague Line, must be the right one because the fishermen of both nations are furious. The eastern third of Georges Bank fell under Canadian control; that tip has valuable swordfishing, offshore lobster, and scallop grounds, all three established fisheries out of Massachusetts and Maine. Howls of resentment on our side. The Canadians wanted to drag and scallop on the American side of the line, saying they had always done so. Howls of resentment from their side.

The day that the decision was announced, the American fishermen appealed to Secretary of State George Schultz for a year's moratorium. Instead they were given two weeks to get their gear out of Canadian waters. The New England congressional delegation promptly called for an International Trade Commission investigation on the extent of federal subsidization of Canadian fishermen. That done, the U.S. put a countervailing duty on Canadian fish entering the country.

Which brings us back to the tariff of 1816. Full circle.

Reinhold Niebuhr and Harry Emerson Fosdick were arguing about the outlook for civilization and reaching no conclusion. Niebuhr then suggested, "If you will be a pessimist with me, decade by decade, I will be an optimist with you, century by century." Fisheries agreements require the perspective of the centuries.

How Big Is Our Garden?

Unitarian Universalist Sermon, June 1997

Dante: "Consider your origins: You were not made that you should live as brutes, but so as to follow virtue and knowledge. Now follow me, for the patterns of the stars are quivering near the horizon now, the north wind's picking up, and farther on there is the cliff's edge we must reach to start down from...."
Divina Commedia, xxvi 118

Schopenhauer: "Thus the task is not so much to see
What no one has yet seen,
But to think what nobody has yet thought
About that which everybody sees."

Saki: Oysters are more beautiful than any religion.... There's nothing in Christianity or Buddhism that quite matches the sympathetic unselfishness of an oyster."

SIXTY-THREE YEARS AGO, on or about this day, I graduated without dishonor from high school, thus confirming my entry into a small college in New Jersey the following September. A classmate named Roger Duncan possessed a Model A touring car with which he kindly deposited me in Christmas Cove for ten days of a solitary life before my parents arrived for the summer. In 1934 we were in the depths of an economic depression (which, as it turned out, grew deeper in 1938). Since it seemed that I would never be gainfully employed and I was bound for a college which taught nothing useful, the time of retreat in Maine set me on a path to a major in philosophy, the science of unanswerable questions, as the least useful field I could think of. And so it turned out.

I retain snippets of Plato, like the Cave and whatnot; I know that Kant had a Categorical Imperative, but not what it was; Nietzsche told me that God was dead but then neglected to prove it. On the other hand, blessings on George Wilhelm Friedrich Hegel, from whom I gathered his simple formula as to how the world works and have used it for decades: Somebody puts forward a thesis, somebody else then brings forth an antithesis or antithesis, and the two battle it out to become a synthesis, which causes an antithesis, and on we go.

When Elaine Pagels applied to graduate school at Harvard University and was asked why she was doing so, she replied that she would like to search for the essential Jesus. Her questioner asked, "What makes you think that there *is* an essential Jesus?" Thesis and antithesis in a nutshell.

So Pagels went off on her search, in the course of which she chased down the development of Satan, who started Old Testament life as a mere messenger and a helpful one at that. He appeared, for instance, on one occasion when a good man was about to lead his donkey over a cliff in the dark, and the messenger stopped him. As antithesis, Satan makes his way into the New Testament as first a fallen angel, next an adversary first of God then of humans, and thereafter a scapegoat for the world's troubles.

To Pagels' undoubtedly immense gratification, about twenty years ago a poor Egyptian farmer was digging up some vegetation, heard his shovel clink on something solid, and exhumed a large earthenware jar containing the Gnostic Gospels. The farmer's mother used a couple of the manuscripts for kindling after he got them back home, until he realized what he had and the unburnt Gospels were preserved for the scholars.

If the Gnostic Gospels had been discovered earlier, the long path to the Quakers and the Unitarians might have been greatly shortened. The Gnostics asked their followers to look inward and find the light within; they recognized females as fully qualified parishioners and as teachers and leaders; consequently, they did not think highly of Paul and felt that his brand of Christianity would be totally unrecognizable to Jesus. Pagels had a new synthesis, ready to be attacked by both the Jesuits and the Christian Right.

Let us now leap to the sixteenth century and to Henry VIII. Henry's thirty-eight-year reign until 1547 was a very complex era of theses and antitheses, some of them interlocking. Martin Luther had nailed his protest on the church door in 1517, the Reverend William Tyndale slipped out of England in 1524 to visit Luther in Worms, and together they arranged for a printer in Cologne to put out an edition of Tyndale's English translation of the Bible. Henry was not about to allow the English peasantry to depart from attending church in Latin, so he sent the Archbishop of Canterbury over to Cologne to buy up and burn as many copies as he could find. A number of the Tyndale Bibles survived, including a few smuggled back into England by Tyndale himself. But his synthesis, alas, was to be brought to ecclesiastical justice. The Archbishop, perhaps seething at the amount of firewood he had to buy in Cologne, cut expenses and had Tyndale strangled at the stake.

In 1534 Henry then broke with the Catholic Church in Rome because he couldn't get a divorce. In bringing the Church of England into being, he expropriated all the Catholic Church land and buildings in Britain, overlooking the fact that the monasteries were the chief producers of honey, hops, and malted barley, thus creating vast shortages of these staples, and much bitterness. As antithesis, the potato, which had been introduced to Europe by Spaniards returning from Peru in 1530, came to England and Ireland at the time of the honey and hops shortages, and became a cheap source of food that set off a quick population explosion.

The King James version of the Bible, published less than eighty years after Tyndale, says, "be fruitful and multiply, and *replenish* the earth, and subdue it...." (italics mine). All the subsequent versions of the Bible have dropped the phrase "replenish the earth." Perhaps the antithesis here is the Newtonian/Enlightenment/Darwinian combo of a mechanistic view of nature.

Let's jump over the centuries to 1997 and the Damariscotta River in search of the essential aquaculture. For nearly a quarter of a century, the thesis ("I'd like an aquaculture lease to grow oysters") found its antithesis in fishermen ("leases are bad for fisheries") and from littoral residents ("an aquaculture operation on the water in front of my house will reduce my property values").

This time, in the Damariscotta, the proposal was for thirty-five acres too shoal for lobstering or crabbing, and never bared at low water to allow clamming. The opposition? From other oyster growers, a new antithesis. A reminder of Afghanistan: we'll join together to fight anybody, but after we conquer them, we'll resume fighting each other. Three hours of testimony on sediment, muddying the waters, interference with moorings which might sometime be placed, the whole works. The unspoken or dimly alluded-to fear? The aquaculture lease would go to a big player. What the other Maine oyster growers need is a new common enemy for a new antithesis. Maybe it will be Connecticut companies with their 30-cent oyster that will make all the little companies band together.

In search of the essential aquaculture or the search for the essential survival, we have to go back to the beginning. Organisms don't experience environments; they create them. Says Louise Young:

> Life altered the atmosphere and gentled the sunlight. It turned the naked rocks into friable soil and clothed them in a variegated mantle of green which captured the energy of our own star for the use of living things. It softened the force of the winds. Life built great reefs which broke the impact of storm-driven waves. Working with amazing strength and endurance, life

transformed a barren landscape and seascape into a benign and beautiful place.

So far, as they say in Vermont.

Life's irresistible urge to be is the prevailing story of the planet. Two billion years ago, life was on the brink of extinction because of a dramatic rise in oxygen, a gas poisonous to then-existing organisms. Oxygen reduced these microbial organisms catastrophically. Then the blue-green bacteria invented a metabolic system requiring the very substance that had been a deadly poison before. They invented breathing as we know and enjoy it.

Now we face a thesis stating that the buildup of carbon dioxide and other greenhouse gases will continue until this planet is not a benign and beautiful place for us. The doomsayers tell us that climate change will move the melting line of the tundra northward and bring an unstoppable progressive release of methane. In Kyoto a world conference will consider the absolute necessity of zoning the entire earth, specifying the requirements of fields and forests for photosynthesis worldwide, because otherwise, in Dr. George Woodwell's words, "The world doesn't work anymore and hasn't since the last doubling of population. I dare not contemplate what will happen by the year 2025."

It is difficult to face such an antithesis. It is even more difficult to contemplate the present thesis. Life is on an unstoppable search for new forms. We cannot halt life; could we not be instead as ingenious and nimble as the blue-green bacteria were 2 billion years ago? Could we not take the boundless need to *be* unto ourselves and could we not give it a welcome ourselves as the essential energy we must draw on for our endeavors to preserve the future, different as it may come to be? "The universe," says South Bristol's own John A. Wheeler, "is something that is looking in at itself."

If you can give an optimistic reply to those questions, then you supply me with my closing quote from a Shakespeare play I have renamed *As You Will Like It*:

> And this our life, exempt from public haunt,
> Finds tongues in trees, books in the running brooks,
> Sermons in stones, and good in everything.

Laws and Kings Can't Save Souls

First appeared in Working Waterfront, *February 1995*

THE LITTLE COUNTRY OF FINLAND, according to London's *Architectural Review*, is currently revising its forestry laws to incorporate "the ethical duty to preserve the soul of trees." The Finns supported this remarkable ethos by building their new embassy in Washington, D.C., on a wooded site so carefully handled that only three trees were taken down to make way for the construction.

The ethical duty to preserve the soul of harbor porpoises is now being honored by the Pinger Project, with at least the statistical hope of saving perhaps a thousand porpoise souls a year from entanglement in gill nets along the Atlantic Coast. The project is an innovative step because historically contentious interests—enviro-types, oceano-Ph.D.s, fisheries regulators, and, most significantly, fishermen—decided to work on the porpoise problem cooperatively.

In a slightly different but just as effective way, the archipelago of Palau, a trust territory taken over by the United States from the United Nations, rose above its U.S.-model government when the old tribal chiefs, many of them expert fishermen, revived the *bul*, an old-fashioned taboo, to close fish spawning areas from April through July. It was touch and go for a while whether the fishermen would obey the *bul*, for the old techniques of hand-lining from outriggers under sail with no refrigerated storage had given way to multihorsepower vessels with big holds and modern equipment. In the old days *bul* violators were beheaded; would a $500 village fine to release confiscated nets and boats be honored? It was.

And a few years ago, Dr. John Todd sailed his "one-and-a-half-ton pickup," a thirty-foot trimaran built by Dick Newick of Maine, to Guyana to determine whether it would be feasible to fish under sail (with a ten-horse outboard to use if the trade winds faded). The best source of fish he found was the bycatch of the Venezuelan shrimpers, who sluiced everything but the shrimp overboard every time they hauled back. If there wasn't much of a sea running, Todd's trimaran could come up alongside and just let the salvaged fish rain down on the netting between the hulls.

Never mind the logbooks and black boxes and quotas and day limits—fishermen are going to fish no matter what and they're going to target the species with a price. That usually means everything else ends up dead on the deck. It's all very well to say it's fertilizer for the crabs, but if there's an ethical duty to preserve the soul of the fish (and that duty has a very practical value for the continuity of the fishery), then the next step is to be able to release the "useless" bycatch alive.

That may take a bit of doing, but if the spirit is there, and there's a growing realization that it is, the solution will be along. As an example, the Nordmore grate caused a bit of fussing and fuming among shrimp fishermen when it first came in, but the fact is that using the Nordmore grate makes separating the shrimp from the rest of the catch easier, while a lot of small flounder are getting a chance to grow. Maybe the cod end of a trawl net can be modified to take the end of an airlift pump and keep the fish up so their swim bladders won't be discombobulated as the net is raised. Who knows? That might not be the solution, perhaps, but it could be something as unpredictable. When it's found we'll have livelier landable fish and rejects that will survive.

We know that fishing pressure has changed the biomass of the Gulf of Maine in ways we don't like; it's the fishermen with the water smarts who can lead the way, now that they know that their livelihood depends on it, to change the balance back to more haddock than dogfish.

Unusual place to find it but here's the operative quote from a play written a couple of centuries ago. It's useful to ponder if you get to thinking that an elected official or some bureaucrat in NMFS is going to solve our problems, instead of ourselves.

> In every government, though terrors reign,
> Though tyrant kings, or tyrant laws restrain,
> How small, of all that human hearts endure,
> That part which laws or kings can cause or cure.
>
> *She Stoops to Conquer,* Oliver Goldsmith, 1770

If the fishermen in Palau or the gillnetters off Portsmouth or the woodsmen of Finland can protect the soul of living things, it's time for us to follow their example.

Reprinted with permission of The Island Institute, Rockland, Maine.

The Grim Reaper's Bycatch

First appeared in Working Waterfront, *October 1999*

MORE THAN 9,000 KIDS under five die each day from illnesses made fatal by malnutrition. Not just the ones you see on the nightly tube with distended bellies and flies around their eyes. Most are in the so-called Third World, of course, but there are plenty in the fully "civilized" worlds of the U.S.A. and Europe.

More than 20 million metric tons of fresh edible fish are discarded every year in the world's commercial fisheries. The U.N. Food and Agriculture Organization (FAO) says that 70 percent of these fish stocks are on the edge of collapse. Those 20 million metric tons equal 44,080,000,000 pounds of fish in a single year. That's ten times the annual seafood consumption of the U.S., or 7.3 pounds for everybody in the world, which passed 6 billion this month. In 1940 world population was 3 billion; now there are almost that many teenagers.

Forty billion pounds of discarded fish every year would provide a lot of cod liver oil, Omega 3s, protein, vitamin A, and a host of other goodies essential to solving the nutrition problem.

The discard problem has been around for a long time. Go to the very second verse of Genesis and you have fishing weather before he even started: "...darkness was upon the face of the deep; and the wind was moving over the face of the waters." Then move to Matthew 13:48: "When it [the net] was full, men drew it ashore and sat down and sorted the good into vessels but threw away the bad."

Interestingly enough, the fishermen we have talked to when there was a 30-pound limit on landed cod scarcely mentioned the damage to their income from the fish they had to discard. Summarized and suitably edited to preserve the decorum of this family newspaper, what they said was simply "this is what we have to do," as they looked at 500 pounds of magnificent cod on deck; the thought of shoveling overboard 470 pounds of perfectly good fish is not part of their fishing philosophy, though they had to do it.

Never mind looking to Congress for a solution to bycatch. Our *Washington Spectator* reports that there are 15,705 private interests who pay for 20,512 lobbyists in D.C.; that's 39 peddlers of influence for each of the 535 congresspeople. The whole U.S. fishery represents maybe 0.01 percent of those private interests. That's the gang that exempted sport utilities from emissions controls because they were "off-road" vehicles, although 96 percent of them never leave the road intentionally. Congress is now irrelevant.

So instead look to a fisherman for a solution.

By golly, there is one, a fellow named Tuck Donnelly, who has been working ingeniously for eight years on what we'd call an intractable problem and he'd call a great opportunity. Up and running since 1993, Northwest Food Strategies has become the biggest protein source for local food banks after the U.S. government, and it's all being done with bycatch. And all under the dynamic leadership of one West Coast pollock fisherman.

Northwest Food Strategies distributed 73,000 pounds of fish in 1993; 1.3 million in 1998; at least 2 million pounds in 1999. It's not nearly the 20 million metric tons of the world fisheries' discards, but after talking with this remarkable fellow, we wouldn't be surprised if he got there eventually.

To get it started, Tuck Donnelly had to take on and convince the National Marine Fisheries Service, seafood processors, and anxious envirogroups who variously suspected that he would be circumventing bycatch and quota strictures (he doesn't), that quality might be dubious (it isn't), and that the bycatch products might leak out and be sold commercially (he set up a watchdog program to be sure they don't). It took years, but now the idea is really taking hold among fishing companies, processors, cold-storage companies, and the federal bureaucrats that all of them can, as Tuck Donnelly says, “take a bite out of hunger,” while taking a big bite out of fishermen’s despair when tossing good fish overboard.

Like the fishermen who looked at the bin full of cod and said, “This is what we have to do,” an idea that came to a pollock fisherman while shoveling salmon over the side has jelled into Northwest Food Strategies—a couple of million pounds of good fish put to use and bringing benefits to consumers and to those making the contributions. All because Tuck Donnelly hung in there. It’s an inspiration for all of us, and one that won’t quit, either.

Reprinted with permission of The Island Institute, Rockland, Maine.

Man or Schmuck?

First appeared in Working Waterfront, *November 1999*

In the year 1604, not many months before George Weymouth landed on Allen Island, the Dutch East India Company hired Hugo Grotius (or Huigh de Groot), a lawyer just turned twenty-one, to write a favorable opinion about who ruled the oceans as part of the company's heated debate with the Portuguese. Like most arguments about the sea, this one had had a long run beginning in 1501 with Vasco da Gama as he gained control of the spice trade for Portugal from the Venetians by blockading the Red Sea. In 1504, to finish breaking the Venetian monopoly, the Lisbon spice market undersold the Venetians by 80 percent. The Dutch were afraid Portugal would repeat this episode with them.

In 1609 Grotius published *Mare Liberum*, a tome predicated on the thesis that fish are the most profitable ocean resource, that the seas should be free to everyone, and that the sea has an endless and inexhaustible supply of fish. To which, three and a half centuries later, Christopher Fry makes reply: "There may always be another reality to make fiction of the one we think we've arrived at."

And indeed, as we all now know, there is another reality. The sea remains, in Grotius's phrase, the common heritage of mankind, but as this century turns, it has also become the common *responsibility* of mankind. Straddling *Mare Liberum* is the equally venerable Public Trust Doctrine, whose provenance includes the city-states of Greece, the emperors of Rome, and the kings of England after the Magna Carta of 1215. English settlers established the doctrine in the Plymouth Colony through the Colonial Ordinances of 1641 to 1647, which in turn became part of Maine law after our state's secession from Massachusetts in 1820, and still survive in full force and effect today. Essentially this doctrine grants a public easement, under complete protection by the State of Maine, allowing any citizen to pass below the high-water mark on the shore for the purposes of fishing, fowling, and navigation, to which purposes the courts have added recreation.

Upholding the Public Trust Doctrine is in the hands of the attorney general, and administered by bureaucrats in the Department of Environmental Protection and the Board of Environmental Protection. Despite the fact that bureaucrats are undoubtedly Nice People, redolent with family values and all that, they are nonetheless unelected burghers commuting to Augusta, whose contact with the sea is limited to sporadic glimpses of the Kennebec between the old Edwards Dam and the river's bend toward Hallowell. As guardians of the doctrine, their job is serious and well done for the most part; their concern is with regulations and obedience to legislation like the Submerged Lands Act, the Natural Resources Protection Act, and such. But they don't have the power to consider the incremental danger of many docks; they do just one at a time.

All this kindles a new doctrine, that of the *Private* Trust: Does everybody have entitlement to a dock with eight feet of water at low tide, whether or not they own a boat drawing six and a half feet?

We know that the world's oceans and especially its estuaries are infinitely fragile, know that humanity has done them more damage in a hundred years than in all the last 4 or 5 billion, can uncomfortably predict that 60 or 70 percent of humanity do and will live within 60 miles of the shore. So when you start to plan a dock or shoreside alteration for your own private convenience, look in a mirror, preferably naked so you can recognize what "poor forked worms" we all are, standing devoid of money, power, access, and lawyers, and mindful of the uncertainties of this life, then ask yourself only a slight revision of the non-rhetorical question: "You have to decide. Are you a man or a schmuck?"

If the latter, never mind.

If the former, skip the private dock, explore alternatives, set an example (maybe a mooring with the dinghy on a haul-off), join the community, fender your rowboat heavily and use the town dock, listen, mellow out and make friends, ask advice and receive wisdom, and come to a careful considerate conclusion: if the town dock is too small, give them a float or two, at a fraction of the cost of the dock you planned—and deductible besides. Make the state's protection of the Public Trust easier. It's a wonderful personal decision, showing respect for the people around you and those to follow.

Reprinted with permission of The Island Institute, Rockland, Maine.

The Trouble with Bob's Rapeseed

First appeared in Working Waterfront, *June 1999*

THE FELL CREEP OF GLOBALIZATION has made its way into our little lives, which are supposed to be, if we can believe Shakespeare's *As You Like It,* "exempt from public haunt." This time around, it's that multinational, Monsanto Company.

On January 30th last, we ventured out to Alna at the top of the tide just to see if there was any detectable salinity in the waters of the Sheepscot River at Head Tide, stepped on a lurking slice of ice, slipped, and broke a femur in a couple of places. We'd lost the hydrometer overboard in the fall, which played hell with the experiment's design and ruined the whole day. Well, not entirely. Earlier we'd been reading the newspaper's *This Day in History* section over morning coffee at the River Cafe, so we were ready for the orienting questions while lying on the fiberglass bed of pain in the ambulance: first our name (we got that right) and then what day it was—"January 30, 1999. Franklin Roosevelt was born on this day in 1882 and Adolf Hitler took over the Reichstag in 1933." The ambulance man turned to the ministering angel with the clipboard and said, "Jeez! Put him down as oriented, in spades." Really enjoyable. You don't often get a chance like that.

Oh yes, Monsanto. The femur recovery stage wasn't all that great, which you might expect for someone in his ninth decade, what with four drift pins, some re-rod, plus a lag-bolt into the hip joint, so we besought a prescription from the orthopod who had inserted all this hardware. Celebrex, he says, only been on the market for a month or so. Just take one after breakfast, and you can go haul all day without feeling any pain in your leg.

So we purchased fifteen Celebrex pills (for sixty-seven bucks) and they do good work on the pain, though they do leave something to be desired with the gut. Then somebody hands us a *Wall Street Journal* story, whose author, after being stonewalled by Robert (just call me Bob) Shapiro, Monsanto's CEO (annual salary $800,000, bonus $2.2 million, value of stock options $2.95 million), learned via the Freedom of Information Act that ten people had died so far from gastrointestinal hemorrhages caused by Celebrex, with so far uncounted numbers suffering lesser catastrophes. One of Bob's spokespeople is quoted saying, "...the drug is performing as expected. The safety profile is what we would expect." Ayup. Just a single chance in 200,000 prescriptions that you die, but if you do, the odds drop to 1:1. So we put in for a refund on the eight unused Celebrex.

Back along before all this, we were cogitating on how to stop erosion so that the thaw and spring rains won't deliver silt to the aquaculture leasehold. The bureaucrats on Prince Edward Island have a recommended list of leafy vegetables, but we don't wish to be tending kale and spinach, so we bethought us of rapeseed (Canadian euphemism: canola). The test plots would show up golden yellow against the undistinguished hay; if any clods with rapeseed went over the bank, they'd be easy to see.

Getting the rapeseed to Maine turned out to be quite a production. We finally found a supplier who would call his wholesaler who would call his distributor who would call his manufacturer's rep, and in a few weeks, we found ourselves possessed of a twenty-five-pound bag of our seed. In the interval we read up some on it, and discovered that Monsanto had done some gene-splicing, so that we could sow the seed without tilling (very convenient) provided that we used a pesticide called Roundup, produced only by Monsanto and bioengineered to kill everything around it but the rapeseed. And the mandatory Roundup is a persistent pesticide that will be in the runoff from the soil for at least three months.

We tried to make our way up the supplier chain, to find someone able or willing to tell us whether or not it was

Monsanto rapeseed. No luck. No labels.

So here we are on the razor-sharp horns of a dilemma: 80 percent of the primary productivity of the world ocean comes from tidal marshes, seagrass and algae beds, and estuaries, all of which are handy by to Clarks Cove; if it's Monsanto seed, it probably won't survive without Roundup, a pesticide we aren't about to use. If it's Monsanto seed, some seeds will be ingested by the local birds and rodents and planted elsewhere, and we don't want to spread gene-spliced seeds. And if it *ain't* Monsanto seed, it's probably not self-tilling. Etc.

And you think you got troubles with 1 15/16 escape vents, mesh sizes, fishing days, what part of Georges is open for what, and all the rest of it? These globalizing days, you can't trust anyone.

Reprinted with permission of The Island Institute, Rockland, Maine.

Local Action Holds Promise for Sustainable Fisheries

Robin Alden

The old adage, "Think globally, act locally," is appropriate not only in a section for Edward Myers's book, but also happens to be one of the areas with the most promise for the management of fisheries.

Today some people in some fishing communities in the Gulf of Maine are envisioning a different fisheries management system. They are acting with the knowledge and the responsibility for living within the limits of the marine resources upon which they depend. At the same time, academic work in the areas of complex systems theory, ecology, and governance is converging on the same truths about the importance of local knowledge and local action. Managing at a local scale may prove to be the way we can actually achieve sustainable fisheries.

What is being envisioned and, in some places, tried at a very small scale is place-based or local area management. It is a management method based on detailed knowledge of local ecology and on a democratic approach whereby users of the resource play a major role in determining rules and enforcement. Place-based management is based on fundamental principals of stewardship and conservation that include limitation of fishermen's ability to be mobile and chase concentrations of fish. It also places limitations on technology to fit the local habitat and behavior of the prey. This approach runs directly counter to the received wisdom for managing fisheries: typically fish and fishermen are regulated across the entire geographic range of the fishery and a few, large-scale mobile fishing units are perceived to be economically efficient. In this model the principal parameters controlled by government are fishing effort and the number of fish caught, rather than when, where, and how fish are caught.

A few assumptions about ecology and governance provide the foundation for all local area management plans:

• Place matters, and local biological differences do exist.

Fish stocks are perceived in an ecological rather than a mathematical manner. Not only do the numbers of fish matter but where they are and what behaviors they are exhibiting (spawning, pre-spawning, feeding, migrating) also matter. The focus on place differentiates this management approach from that traditionally used by fishery scientists and state and federal managers. Fisheries science, involved as it is with the analysis of large population units, has not focused on local level phenomena, such as the changes in behavior and distribution of local populations associated with the collapse of a stock that are so often described by fishermen.

The local area management approach embraces recent research by ecologists, fish behaviorists, and work that incorporates historical fishermen's knowledge into the scientific literature. The combination of ecological research and fishermen's knowledge and observation are providing growing evidence that suggests complex, localized phenomena within previously viewed homogeneous stocks of commercial species. As an example, scientists and fishermen alike are beginning to agree that cod may return to their natal spawning grounds to spawn, just as salmon do. Cod have elaborate mating rituals and perform them only on specific types of substrate. Fishermen have reported for years different "batches" of fish at different times of year with recognizable physical differences. Over 270 specific fishing grounds in the Gulf of Maine have been identified from historical fishing records, each named and described by fishermen over the last 120 years.

It can be hard for the non-fishing public to understand the existence of local habitats in the ocean. For most, the marine environment is out of sight and mostly out of mind. Although fishermen are aware of how varied the bottom types are, and of the ways that currents operate in specific locations, to the general public, the ocean is one big, homogeneous bathtub, with a smooth bottom. Few people think of the ocean as having specific locales within it.

It is easy to understand that place matters with softshell clams. It is easy to see that clams don't move out of a clam flat. It is less easy to know how important local spawning is to recruitment on that flat: does the clam spat produced by those clams stay put and settle on that same flat? When one jumps to cod, haddock, and flounder, however, even the questions are not clear. Groundfish management typically has assumed that one or two homogeneous stocks of fish, such as codfish, mix freely over large geographical areas and have no important local biology.

These new scientific perceptions have profound implications for management. For example, since knowledgeable fishermen go where the fish are, they may, in fact, be moving from one biologically distinct group of fish to another, hitting each grouping when it aggregates before spawning, so that the entire effort of the fleet is actually applied on each smaller stock. This way of fishing inevitably results in successive depletion of each stock and the eventual loss of reproductive capacity within the entire population.

- Complexity and uncertainty.

The marine environment is highly complex and fisheries management always operates with a high degree of uncertainty. It is very difficult to be sure of cause-and-effect in the marine environment because of its complexity and the difficulty of long-term study. Complexity theory tells us that the rational behavior in the face of uncertainty is adaptive behavior, which is based on many small decisions from which one can learn rather than one big irrevocable decision and action. The corporate world has begun to embrace this adaptive approach.

The second element of all conservation is governance. Those who govern must promote the conditions for stewardship. New Englanders are culturally trained to understand the value of participatory democracy. Now political scientists and economists studying natural resource management around the world in contexts ranging from fisheries to irrigation rights have developed a body of work that emphasizes decentralized, adaptive decision making as the optimal approach to deal with natural complexity. Called co-management, this approach allows local, decentralized groups to have management authority and to exercise it through participation in decision making.

- Fishermen take responsibility for conservation.

The usual model, which we now understand doesn't work, has government giving orders and fishermen pushing back, doing as little conservation as possible. In this model, responsibility for conservation lies with government alone. In contrast, local area management places responsibility with fishermen. It adds an additional, responsible, functional level of organization and action at the local level, while still keeping management within a broader state, regional, and federal context.

At first blush the idea of fishermen being responsible for conservation sounds backwards--like the fox guarding the henhouse. However, there is actually no other option. It is fishermen who carry out conservation rules on the water. No state has the money or equipment to be capable of enforcing rules that are not internalized by those running the boats or pushing the clam hoes. Maine's lobster fishery has a long history of self-enforcement of common-sense fishery rules such as the protection of breeding females and juveniles. It is the lobsterman, not the government, who throws marketable female lobsters over the side so they can breed again.

What does it take to establish the conditions for effective local stewardship of resources? The answers lie in establishing local-scale institutions in which those who know the marine area and the people fishing can participate. They involve requiring certain base knowledge and responsibility as a condition of participation in fishing, similar to the existing lobstering apprenticeship requirements. And they require the evolution of scientific institutions that respect the observations of fishermen and collaborate with them to explore scientific questions. Happily, this evolution is underway both in Maine and in New England.

It is encouraging to see the informal, spontaneous emergence of small groups cutting their teeth on this type of participatory management. Several years ago the Cobscook Bay scallopers organized to ask the Maine legislature for laws to limit the scale of the fishery in the bay so that the local fishermen could stretch the annual scallop harvest over a longer time, rather than having it eliminated within a week of opening day by a huge pulse of effort. And in Canada and Maine small-scale groundfishermen and those who have lost access to the fishery through federal regulation have met, organized, and articulated proposals for sustainable groundfish harvest.

These groups communicate with each other through various mechanisms such as Northwest Atlantic Marine Alliance and Saltwater Network. They are deepening their understanding of the daunting community-organizing task necessary to enable fishermen and their communities to make good choices about the use of their resources.

In Maine place-based management is being tried with three marine resources: shellfish, lobster, and groundfish. Maine law allows towns to manage softshell clams through town shellfish committees if they meet certain standards. In some towns committees made up of diggers, town officials, and the general public participate in setting rules for access to the flats, opening and closing of flats for conservation purposes, and enhancement projects such as moving seed clams. The towns are also responsible for enforcing their own rules. Importantly, the clam committees have the authority to limit the number of licensees allowed to dig the town flats.

Maine also has a unique lobster management system that is built on the traditional, informal system of lobster territories. The Maine Coast is divided into seven zones, each with about the same number of lobstermen. Each lobsterman is entitled to vote for one representative from his or her geographical area within that zone. Those representatives then make up the zone council. Each zone council votes on major issues for the zone: the number of traps each lobsterman is allowed to fish, when to fish, and how to rig the traps. The zone council also may vote on the ratio of new lobstermen entering the fishery in their zone to existing lobstermen leaving the fishery, known as the entry/exit ratio for their area. This provision allows the lobstermen themselves to control the increase in fishing pressure in their zone.

Finally, a group of community-based fishing and conservation groups that operate through the Northwest Atlantic Marine Alliance devised an area management plan for Gulf of Maine groundfish as part of recent discussions on Amendment 13 to the New England Fishery Management Council's groundfish plan. The Council's plan governs all aspects of fishing for groundfish, particularly cod, haddock, and yellow flounder, in the Gulf of Maine.

Although their effort was rejected by the Council, the Northwest Atlantic Marine Alliance plan stands as an excellent example of an attempt at local area management in a complex, multi-species fishery that involves many gear types and vessel sizes. The plan called for extra protection of the nearshore areas of the Gulf of Maine where cod and other groundfish spawn and group-up in pre-spawning aggregations. Their proposal included strict limits on the types and scale of fishing technology in the inshore critical habitat areas and most importantly, required that fishermen choose one area and operate within it for at least five years. By limiting mobility the proposed plan would have made it impossible for fishermen to pulse-fish localized concentrations of fish. The ideas behind the rejected proposal were eloquently described and supported in a letter signed onto by more than eighty scientists from around the world.

Many groups such as the Northwest Atlantic Marine Alliance operate based on clearly stated ecological principles and principles of practice. These principles provide a foundation of common belief that make it possible for the members to share their knowledge, disagree constructively, and force themselves to think about the ecological and community goals, rather than their own personal gain. When these place-based groups reach a decision, the result truly can be powerful within the larger political arena precisely because the decision is based on something larger than individual interest and is for the long-term.

The Gulf of Maine is a phenomenally productive system. In the last fifty years we have learned much about its ecology and about the process of human decision making and governance. Most of all, we have learned that we need to keep on perfecting that knowledge, adapting to the changes in a dynamic natural system and our own human system. Through local action we have an opportunity to sit lightly on that system, achieve the grail of sustainable fisheries, and share in its rewards for generations.

ROBIN ALDEN was Maine Commissioner of Marine Resources from 1995–97. Previously she was publisher and editor of *Commercial Fisheries News,* a trade newspaper which she founded in 1973. Currently Ms. Alden is working on fisheries governance and cooperative research with community-based groups in Maine and Atlantic Canada. She is developing the nonprofit Penobscot East Resource Center in her hometown of Stonington, Maine.

FOUR

SPIRIT

The Holy and the World

Union Church Sermon, May 1987

MY TEXT ON THIS SPRING DAY—close to that day in May when we are inclined to shout at Nature, "Stop! Stay just the way you are, before the weeds, before the cutworms get at the broccoli. Don't change a thing!"—is from that prophet whose name, with a small *j*, moved into the common language as jeremiad.

"Thus spake the Lord: 'I brought you into a fruitful land to enjoy its fruit and the goodness of it; but when you came in you defiled my land, and made the home I gave you loathsome.'"

Wendell Berry, that remarkable Kentucky farmer, writes that "perhaps the great disaster of human history is one that happened to or within religion: that is, the conceptual division between the holy and the world."

How did this accusation come about? The Roman Empire, with many mouths to feed, sensibly chose North Africa as its granary, but now a quarter of a million acres of land is lost to the Sahara Desert every year.

The Sahara situation is just a little splinter. Around the world, 200 times that amount of land—50 million acres—changes from cropland to desert every year.

A billion people, about a fifth of the total population, has no access to potable water. More than 2 billion people have never even seen a faucet. The United States pumps 122 billion cubic meters of groundwater a year (a cubic meter holds 263 gallons) and is therefore using 380 gallons per day for each one of us. Of that, one-fifth or 26 billion cubic meters is unrenewable. I am delighted to learn that Maine's Department of Environmental Protection is requiring that a developer prove the groundwater supply for each of thirty lots laid out is adequate before being able to proceed with building.

The United States is not alone, of course. The Soviets are worried, particularly along the southern tier of their republics, when supplies around the Aral Sea begin to run short.

A hundred and twenty years ago, John Wesley Powell, the surveyor-cum-geologist who took the first wilderness rafting trip down the Colorado River, reported to Congress that the development of the western lands he traveled over would be severely limited by water supply. He was laughed off the floor of Congress. Within fifty years the city of Los Angeles was casting about for places to borrow water. Nowadays, New Mexico and Arizona have joined California in crisis; they are mining water and fighting over it, while the land itself becomes salinated and drops fifteen or twenty feet as groundwater is withdrawn.

Look closely at your can of frozen orange juice concentrate. In the fine print it says, "...from oranges grown in Florida or Brazil." Tropical forests, the self-regulating rain forests that take seven or eight thousand years to create and stabilize, are disappearing at the rate of 27 million acres a year. Continue to do this and you get a major loss of oxygen, a major rise in carbon dioxide, bringing weather changes and accelerating the greenhouse effect.

Dr. Frank Notestein, who lived down the street from us during my exile in New Jersey, categorizes the development programs of countries in three stages. First is slow growth, with both birth and death rates high; the second is fast growth, when improved medicine and sanitary conditions let birth rates go up and death rates go down; the third is more or less stability, where economic and social gains hold the birth rate down. Buckminster Fuller suggested a novel solution for India's overcrowding: spread electricity everywhere and give every household a television.

There are dozens of developing countries wherein 40 percent of the population is under fifteen. In Nicaragua, which is having a little trouble developing these days, 50 percent of the population is under fifteen. Right now the world is chewing up 40 percent of available photosynthesis—forest, grass, cropland, and aquifers—to support the human population. If you've ever had any fifteen-year-old boys, you know how they can eat. At the turn of the century there will only be half as much cropland left to support double the population of the planet.

It will take more than computers. The United States Air Force has unlimited access to computers and is unable to locate 31 percent of its inventory of spare parts and ammunition.

How did this come about? One of those councils at Nicaea, the one in a.d. 787, asserted that Jesus was one with God, and that both in one so loved the world. Dostoyevski had one of the Brothers Karamazov dream about the wedding in Cana, where Christ's first miracle was performed. The man wakes up from his dream joyously, throws himself down on the earth, and embraces it. He kisses the earth, forgives the earth, begs its forgiveness, and vows to love it forever. No division there.

John Stewart Collis makes a case for polytheism in thinking about the conceptual division between the holy and the world, which Dostoyevski just solved in his own way. "Whereas under polytheism the gods were intimately connected with the earth and stimulated veneration for it, under monotheism deity was extracted from the earth. God was promoted to higher regions. He went completely out of sight. It became possible to fear God without fearing Nature—nay to love God and to hate his creations."

Let me return to Wendell Berry:

> If God is not in the world, then obviously the world is a thing of inferior importance, or of no importance at all. Those who are disposed to exploit it are free to do so. And this split in public attitudes is inevitably mirrored in the lives of individuals: A man can aspire to heaven with his heart and mind while destroying the earth and his fellow man with his hands, and do his

> living in the next world. Which completely ignores the fact that the here is antecedent to the hereafter, and that indeed the Gospels—and I would add, the meaning of the table before us—make one's fate in the hereafter dependent upon one's behavior here.

So my essay comes down against this conceptual division. On this promising day of rebirth all over the place, we in this little island church are fortunate in being able to see that there is no division, conceptual or otherwise, between God and Nature, anguished as the world sometimes and somewhere is.

Hence, in conclusion, two more snippets. From our late Nobleboro neighbor, Elizabeth Coatsworth:

> If Americans are really to become at home in the world, it must be through the devotion of many people to many small, deeply loved places, the field by the sea, the single mountain seen from a door, the lush and tropical jungle, the island of trees and farm buildings in the wheat, must be sung and painted and praised until each takes on the gentleness of the thing long loved, and becomes an unconscious part of us and we of it.

And of course from J. R. R. Tolkien:

> The rule of no realm is mine, but all worthy things that are in peril as the world now stands, those are my care. And for my part, I shall not wholly fail in my task if anything passes through this night that can still grow fair and bear fruit and flower again in days to come.

I think that this is part of God's plan.

A final snippet, the closing words of Sir Thomas Browne's *Religio Medici,* 1635, treated as prayer:

> Bless us in this life with but peace of our consciences, command of our affections, the love of Thy self and our dearest friends.... These are, O Lord, the humble desires of our most reasonable ambitions and all we dare call happiness on earth, wherein we set no limit to Thy hand or Providence. Dispose of us according to the wisdom of Thy pleasure; Thy will be done, even though in our own undoing.

Edward O. Wilson's *The Future of Life*

Unitarian Universalist Sermon, March 2002

EARLIER THIS MONTH we were bound for Rockland on some errand or other and decided to count the number of "God Bless America" signs on the way. The last one before our destination was at a restaurant with one of those plastic signboards to which letters are attached one at a time. The wind or the rare rain had removed one letter, so the sign read, "Go— Bless America." Having zero conviction that God should bless America any more than she already has, or any more than she might bless Tasmania or Tierra del Fuego, the sentiment "Go— Bless America," with its faint whiff of Isaiah and the Gospel instruction "go forth to all the nations," seems to us much better than Kate Smith's musical version.

One person who is doing this, and in a wondrous way, is Edward O. Wilson, research professor and entomological curator of Harvard's Museum of Comparative Zoology, in his book, *The Future of Life.*

The book starts its work with the dust jacket. What looks like a beautiful arrangement of flowers and animals and seeds is an assemblage of sixty-one species which are either extinct or about to be. Then, eight pages past the flyleaf, is a concordance diagram of the species plus a two-page list of the sixty-one disappearances. It's really poignant. Dr. Wilson bemoans the loss of each one.

Everybody in the world should bemoan the loss of a single species, whether it buzzes, bites, or is simply bizarre, and Dr. Wilson makes a vital case for the preservation of every one, known or presently unknown. Fewer than two million species are in the official register, with Latin names supplied by Linnaeus or other taxonomists; of those, fewer than twenty thousand have been investigated much beyond naming them and proving that they are a distinct species. Nobody knows how many species have not yet been identified. The educated guess is a very cautious one, somewhere between five million and a hundred million, an echo of Emerson's dictum that, "a weed is a flower whose virtues have not yet been discovered."

Dr. Wilson would have us begin by concentrating on twenty-five hotspots around the world which are less than 1.5 percent of the globe's area but contain 44 percent of the plant species in the world—and contain them exclusively—and more than a third of all birds, mammals, reptiles, and amphibians. Fifteen of the hotspots are tropical rainforests: Brazil, of course, Central America, the tropical Andes, the Greater Antilles, West Africa, Madagascar, the Western Ghats, Myanmar, Indonesia, the Philippines, and New Caledonia. The wilderness hotspots, or frontier forests, are Amazonia, the Congo Basin, New Guinea, Canada, and Russia (the forests of Malaysia, Sumatra, and Borneo essentially are gone). On the marine side, Dr. Wilson would have us preserve coral reefs, establish Marine Protection Reserves, and otherwise care for the oceans.

Except for Canada, none of the places is in our ready consciousness. The seamless fabric of biodiversity ought to be, and it is very positively in our interest to preserve all species.

As a prime example, look at Amazonia, where tribal medicine arises from experience with fifty thousand native species of flowering plants. Listen to a partial list of illnesses and their treatment:

- Motelo sanango (snakebite and fever)—*dye plant*
- Anemia and conjunctivitis—*monkey ladder*
- Amoebic dysentery—*caimito*
- Mouth sores and fungal infections—*toad vine*
- Tumors—*renaquilla*
- Rheumatism—*calabash*
- Toothache—*milk tree*
- Skin inflammation—*dragon's blood*
- Hemorrhaging—*fer-de-lance plant*

- Snakebite—*swampimort*
- Malaria—*wild mango*

In the U.S.A. 40 percent of prescription drugs come from plants, organisms, and animals, with nine of the ten leading drugs from organisms, for a total annual revenue of $20 billion. Of antibiotics, 85 percent come from organisms, yet only 2 percent of the world's ascomycetes species produce them.

Dr. Wilson develops careful examples on the values hidden in newly discovered species. One example is an extremophile bacteria living in the boiling springs of Yellowstone. This microscopic beast produced a heat-resistant enzyme essential to the synthesizing of DNA. CETUS, the company that discovered it, is making $200 million a year from the bacteria for its ability to accelerate genetic mapping.

A couple of economic botanists proved that single harvests of plants in Belize yielded $290 on one acre and $1,330 on another. Other analyses showed a value of $91 in Guatemala and $135 in Brazil for medicinal plants, and an additional $1,275 for tropical pine. Per acre. All these figures are for sustainable levels of harvest and are much higher than figures for ranching or slash-and-burn agriculture, both of which are usually unsustainable.

The last chapter of *The Future of Life* is entitled "The Solution." It contains a litany of direct and essentially simple suggestions for saving the biodiversity of the planet and thereby saving its human occupants. At the outer edges of the author's hope lies an even split: 50 percent for biodiversity and 50 percent for humanity.

In this momentarily broken world what the solution comes down to is re-energizing the entire planet. The people with the money to do this are the oil companies and the investment bankers, the large foundations, the 200 corporations who control 80 percent of the wealth in the world. The first step is to get rid of our widespread denial that the earth is being destroyed. Then follows the second step: realization that you can't have a good economy unless you have a good ecology. The third step is one of assessing value: however slippery the figure, currently the thoughtful estimate is that biodiversity is worth about 33 trillion dollars per year, approximately ten times the gross world product.

The fourth step is acceptance of this ethical and spiritual challenge by those most fiscally able to do so, and their realization that meeting the challenge will be eminently worth doing, much more so than a perpetual dance with sheiks, warlords, lobbyists, and the grotesque subsidies of the military. It would be a brand new economic game with a wonderful goal.

A quote to end with, which I cannot identify:

> People will try to tell you that love is a luxury and that life in all its miraculous beauty is less urgent right now than the need to retaliate against the forces of evil. I am here to tell you that unless we respond with love, we will certainly hand evil its great and final victory. Go out, right now, and plant yourself in the middle of that which you love most, the thing within you that is most alive.

Now listen carefully, because as that love cracks your heart open, it will tell you exactly what this broken world needs from you. This is your holy work, and it cannot wait. Make it big this time. Make it so.

Shepherds and Shepherdesses

Second Congregational Church Sermon,
Newcastle, Maine, May 1984

SOMEWHERE IN HIS BOOK *The Warden,* Anthony Trollope wrote, "One may not be said to have survived an English winter until the second week in May." This may apply to a New English winter, too, but we have endured, nay survived; it is the fourth Sunday of Easter, it is well into lambing time, and the time also, as the latest newsletter suggests, for shepherds and shepherdesses.

I have always assumed that the passage in the sixth chapter of Acts was the basis for having both deacons and trustees in a Congregational church. The very name comes from the Latin *gregs gregis* meaning a herd or flock and a *congregs* is a flocking together. Thus deacons and trustees serve in their particular ways and do their serving together.

Then the sixteenth verse of 10 John—"and there shall be one fold"—specifies the size and scope of the shepherds' and shepherdesses' job, which is the whole world, and that is a scary assignment.

Those of you who have had to listen to me before know of my fascination with the smallness and fragility of the world, and the narrow range in which human life can survive unassisted by mechanical devices, from a few feet of water to about ten thousand feet into the atmosphere. At this very moment, the little planet we're riding on is flying on its path around the sun at 56,299 miles per hour. The whole solar system on which we are piggybacking is itself hurtling along at 38,506 miles per hour on some course or other, perhaps in a circular orbit or in an elliptical one, but circling what? Then, of course, in this latitude we are spinning at 460 miles per hour. All told in the three different directions, we'll have traveled 95,265 miles during this service. So will the 4.5 billion people riding with us, all depending on the fact that this terrific speed through space won't leave behind the slender cover of life-giving atmosphere.

Oh, relax, a scientist tells me. There's nothing in space, so there's no friction. Ah, I reply, but if there's nothing in space, how come there's something like the planet earth and all the other heavenly bodies? The scientist is mute and we are left with the miracle, along with the responsibility for the billions of fellow passengers, beginning their lives, living them, and completing them in an endless procession, all needing shepherds and shepherdesses.

With all these big numbers, I am striving to understand computers and what they can do. I was reading a book called *Computer Power* by Professor Joseph Weizenbaum of MIT and stumbled upon a surprise ending:

> For the present dilemma, the operative rule is that the salvation of the world depends on converting others to sound ideas. That rule is false. The salvation of the world depends only on the individual whose world it is. At the very least, every individual must act as if the whole future of the world, of humanity itself, depends upon him or upon her. Anything less is a shirking of responsibility, and is itself a dehumanizing force; for anything less encourages the individual to look upon himself as a mere actor in a drama written by anonymous scribblers, as less than a whole person, and that is the beginning of passivity and aimlessness.

One shepherd. One shepherdess. Me. You.

A little gloomier at the outset is Wendell Berry in his *Recollected Essays*—this one is entitled "Discipline and Hope":

> The condition in this country now is one in which the means or the discipline necessary to the achievement

> of professed ends have been devalued or corrupted or abandoned altogether. We are asked repeatedly by our elected officials to console ourselves with that most degenerate of political arguments: Though we are not doing as well as we might, we could do worse, and we are doing better than some. We are offered peace without forbearance or tolerance or love, security without effort and without standards, freedom without risk or responsibility, abundance without thrift.

But if that is what we are being offered, we do not have to accept the offer. The good shepherd may instead listen to Paul that tribulation worketh patience, patience worketh experience, and experience worketh hope. And hope is a virtue.

There is the notion that troubling times may be what Asians call the "withdrawal of the mandate of heaven." This situation arises whenever an administration or regime loses the ability or desire to share with the people a sense of what is needed and what is just, and the capability to rule in accordance with those sensings. No one in particular ever says that the mandate is revoked; this feeling just gradually filters through the people until it colors the overall climate of opinion. People feel that their own capabilities and the exercise of their imaginations or their abilities to serve directly are ignored or impeded by the large institutional structures of the society. While people's capacities are increasing, frustrating rigidities appear which may reduce their opportunities to serve. This situation, again, need not be accepted. The mandate of heaven cannot be withdrawn if it is held to by the individual shepherds and shepherdesses.

Holding to this *mandate* is a visible illustration of what William James called, "Those tiny, invisible molecular moral forces which work from individual to individual, creeping through the crannies of the world like so many rootlets, or like a capillary oozing water, yet which, if you give them time, will rend the hardest monuments of men's pride."

"Joy—human joy," said Teilhard de Chardin, "is the infallible evidence of the existence of God." Let me return first to the computer man's phrase that the salvation of the world, the human world, depends only upon the individual; then let me borrow some joy in paying homage to the shepherds and shepherdesses of this church, and let me do so with Dag Hammarskjöld's quatrain:

> Give us a pure heart, that we may see thee;
> A humble heart, that we may hear thee;
> A heart of love, that we may serve thee;
> A heart of faith, that we may live thee.

How to Enjoy an Election Year

4 October 1992 (unpublished)

On Friday August 14 last, I was on board the 30-foot sloop *Cassiopeia* preparing for a passage to Halifax. Lying to a mooring at sunset in Frenchboro Long Island, seven or eight miles south of Cadillac Mountain, my son-in-law set up a forestay to turn the yacht into a cutter, just a precaution against heavy weather. To me it seemed too cautious. As we retired in the last of the light, I had a clear vision of picking up the Overfalls whistle buoy off Cape Sable in the rosy dawn of Sunday as we reached along in a southerly, signaling for a left turn as we set the spinnaker for a day's run up the Nova Scotia coast. Filled with scrambled eggs and sausages plus four English muffins with peanut butter and strawberry jam and a fifth of coffee, I imagined impressing my son-in-law with my seamanship and general handiness plus spryness belying my advancing years.

We made the Overfalls buoy right on the button at 6:00 A.M. Sunday morning, twenty-four and a half hours out of Frenchboro. But there was no sunrise. No breakfast. We were close-hauled in an ungodly breeze a list east of south; there were three reefs in the mainsail and one in the staysail; the Fundy tide was falling across the wind; I was helplessly seasick and absolutely, irretrievably good for nothing. The only thing handy about me, as I lay on the cabin sole, was that my head was handy to the head.

My son-in-law is an RN and an EMT and while not exactly a Florence Nightingale type, has a quality of mercy which he put into high gear. Before the buoy was even out of sight in the fog and rain, he abandoned Halifax as an objective and bore off for Yarmouth. As a result, the confused seas, instead of coming green over the bow, came aboard in dollops over the quarter and I could abandon my determination to die. We didn't have a chart for Yarmouth, but I had borrowed Roger Duncan's copy of a cruising guide to the Nova Scotia coast, so we used that to get in to a snug berth about 150 yards away from the Yarmouth ferry dock. As a matter of fact, we waited at the outer bell for the *Bluenose* from Bar Harbor to show us the way in.

After a recovering day or two of fog and rain, we had some lovely sailing to Meteghan and Petit Passage and then to Digby, where I caught a bus to return to Yarmouth, boarded the *Bluenose,* and thence home.

We were so busy having fun that we listened to no radio except the Coast Guard's Channel 16 and stayed far from newspapers or television. Not until I strolled aboard the *Bluenose* that Friday afternoon did I realize that I had totally missed the Republican Convention. And I also realized what a wonderful situation that was. I was fully rested and full of good food; my mind was as clear as it ever gets these days. I decided to keep it that way. At the ship's rail I had a chat with a Japanese graduate student from Castine; other than that, I was content with a drugstore penny shocker that cost $4.95 Canadian and with my own thoughts until the red flashing bell came up off Schoodic Point.

Back in civilization, how could I sustain this happy state of mind without escapism? Surely not with the nightly ministrations of Dan or Peter or Robert or Jim [Rather, Jennings, MacNeil, Lehrer]. Probably not even with Gibbon's *Decline and Fall*—too many morose parallels. Then I recalled a snippet of a quote—"...if anyone asks me what I think the chief cause of the extraordinary prosperity and growing power of this American nation, I should answer that it is due to the superiority of their women"—and I said, that's the fellow who will take me blithely through to Inauguration Day! Alexis Charles Henri Maurice Clorel de Tocqueville, author of *Democracy in America,* four fairly slender volumes published in 1835 to

1840, re-issued in a single-volume translation in 1966. Anybody who could get off a sound bite like that in 1835 deserved further study.

An aristocrat in the Chamber of Deputies who saw trouble ahead in the French Revolution of July 1830, he thought it best to get out of Paris until it blew over. He applied for leave to go to the United States to study prison reform, an application not granted until 1831. A scholarly colleague named Beaumont, also uncomfortable with revolution, applied for leave to study U.S. slavery, so off they went together.

De Tocqueville didn't spend much time on prison reform and Beaumont didn't spend much on slavery, but they investigated just about everything else.

Now I am not going to give you a book review, but I do commend to you some of the heady delights of a book which has had such tremendous effect on the politics of many countries. How about this as a topic for Columbus Day next weekend:

> When the Europeans landed on the shores of the West Indies, they thought themselves transported to the fabled lands of the poets. The sea sparkled with the fires of the tropics. Here and there little scented islands float like baskets of flowers on the calm sea. Everything seen in these enchanted islands seems devised to meet man's needs or serve his pleasures.
>
> Death lay concealed beneath this brilliant cloak, but it was not noticed then, and moreover there prevailed in the air of these climates some enervating influence which made men think only of the present, careless of the future.
>
> North America seemed very different; everything there was grave and serious and solemn; one might say that it had been created to be the domain of the intelligence, as the other was that of the senses.

What a nice way to compose the differences of opinion about Columbus who, in an egregious bit of poor scheduling, discovered the West Indies precisely five centuries before an election year. And isn't de Tocqueville's capsule of wisdom worth more than whatever *MacNeil Lehrer* and *Washington Week in Review* will say about Columbus next Friday [officially the 500th anniversary of Columbus's discovery]?

Beaumont and de Tocqueville were in the United States from April 1831 to February 1832, between the election of Andrew Jackson in 1828 and his reelection in November of 1832. Listen to de Tocqueville, who incidentally had his twenty-sixth birthday in the wilderness settlement of Saginaw.

Chapter heading: "The Presidential election may be considered a moment of national crisis"—

> Up to the present, the political circumstances of the national election time have presented no real danger. Nevertheless, the time of the Presidential election is a moment of national crisis.... Parties feel the need to rally around one man in order more easily to make themselves understood by the crowd. Generally, they use the Presidential candidate's name as a symbol; in him they personify their theories. Hence the parties have a great interest in winning the election to show, by his election, that their doctrines have gained a majority.... Long before the appointed day arrives, the election becomes the greatest and one might say, only affair occupying men's minds. The factions redouble their ardor; then every forced passion imaginable spreads excitement in broad daylight.... As the election draws near, intrigues grow more active and agitation more lively. The whole nation gets into a feverish state; the election is the daily theme in newspapers and in private conversation, and the sole interest.

Alexis was also looking toward Jackson's run for reelection in 1832.

Chapter heading: "When the head of the executive power is reeligible, it is the state itself which intrigues and corrupts; the desire to be reelected dominates all the thoughts of the President of the United States"—

> When a simple candidate forces himself forward, his maneuvers can only take place in a restricted sphere. But when the head of state is in the lists, he can borrow all the power of the government for his private use. It tends to degrade the political morality of the nation and to substitute craft for patriotism....
>
> The founding fathers agreed that, besides the people, there must be a certain number of authorities which, though not entirely independent of it, nevertheless enjoyed within their sphere a fairly wide degree of freedom; though forced to obey the permanent directions of the majority, they could still struggle against its caprices and refuse to be tools of its dangerous exigencies. With this object, they concentrated the whole executive power in one man; they gave wide prerogatives to the President and armed him with the veto to resist the encroachments of the legislature.
>
> But by introducing the principle of reelection they destroyed a part of their work. They gave the President much power, but took away from him the will to use it....
>
> Reeligible, the President of the United States is only a docile instrument in the hands of the majority. He loves what it loves and hates what it hates...anticipating its complaint and bending to its slightest wishes; the lawgivers wished him to guide it, but it is he who follows.
>
> In this way, intending not to deprive the state of

> one man's talents, they have rendered those talents almost useless; to preserve resources against extraordinary eventualities, they have exposed the country to dangers every day.

Alexis de Tocqueville had his critics—one of the more memorable if captious being Saint Beuve, who said, "What a pity that he began to think before learning anything"—but he can't be dismissed as just an author with a single bestseller. The timing of his American visit was right on the wave of Jacksonian reforms, which culminated in the peak of laissez-faire capitalism a half century later, and humanitarian reform movements focusing on prison, education, women's and children's rights, banking, and so on.

Alexis de Tocqueville lived from 1805 to 1859, John Stuart Mill from 1806 to 1873, Thomas Carlyle from 1795 to 1881, and Karl Marx from 1818 to 1883. Ralph Waldo Emerson lived from 1803 to 1882. De Tocqueville corresponded with Mill, and even wrote an article on democracy for him. Carlyle's *French Revolution* appeared in 1837 and de Tocqueville's reading of it helped him predict the shape and size of the French Revolution of 1848. Although no letters have been preserved, some think that Marx and de Tocqueville were in correspondence after the appearance of Marx's *Manifesto* in 1848. In the largest sense, the political choice of that century was between the ideas of Marx and de Tocqueville.

After serving a brief term as foreign minister of France in 1849, de Tocqueville returned to delving into the antecedents of the French Revolution, much deeper in fact than Carlyle, and published *The Ancien Regime and the Revolution.* Illness and death overtook him as he was preparing to go to work on a second volume.

There is an old wheeze which I tell about that man of all work and very little play, Eben Fossett, who made a weekend trip one time from Bristol to Boston. On his return to work

Monday morning, his fellow workers asked him what sights he saw. "Godfrey mighty," he replied, "there was so much going on at the depot I never did get uptown."

I feel much the same way. There was so much going on in 1835 that I just haven't time to bother with this year's election. And it's just wonderful.

Marshalling Diversity

Unitarian Universalist Sermon, Undated

DIVERSITY. FOR MANY REASONS it's a word that's become fashionable and there are lots of instructions that you'd better tolerate it. It surprises me that there's such a fuss about it, as here we are faced with the miracle that every individual is different. I look out upon a sailor-headmaster-novelist-historian, a limnologist worried about the population explosion, a musician, an accomplished actress, an expert on quilts, a public utilities executive, and heaven knows what all. I find no difficulty in toleration.

So I have decided that if there is all this tolerance going around, it should be possible to find the toughest diversity extant and discover just what my tolerance toward it was. After some thought, I selected Iraq. I don't expect many people to agree with me, but I appeal for your tolerance.

There's a very practical reason for choosing Iraq: the government is tyrannical and bloody and dictatorial and does not like America very much. This is a recipe for terrorism and, despite the pious representations of anti-terrorist groups in the CIA, FBI, State Department, National Security Council, military counter-intelligence offices, and so on, there really isn't much you can do about terrorism except make the terrorists change from adversaries into friends. If they're not friendly, they can mount such grisly operations as the bombing of the World Trade Center.

As I have said previously, miniaturization now makes it possible to stash a device with the power of the Hiroshima bomb in a locker at the 40th Street bus station or the baggage room at Paddington Station. I am interested in the continued safety of my children and grandchildren, who carry their passports in the hip pockets of their jeans and are likely to be found anywhere. When someone suggests to me that we have a truly mobile society, I do not have to look very far.

Now in looking for guidance as to how to proceed in the toughest project imaginable, that is, recovering, if indeed we ever had it, the amity of the Iraqi people, we can look to a statesman like Edmund Burke who said, "The use of force alone is but temporary. It may subdue for a moment but it does not remove the necessity of subduing again, and a nation is not governed which is perpetually to be conquered." Or listen to a rather cynical critic like Jacques Anatole François, "To disarm the strong and arm the weak would be to change the social order which it is my job to preserve. Justice is the means by which established injustices are sanctioned." Or contemplate poet John Milton's words in *Paradise Lost,* "Who overcomes by force, hath overcome but half his foe."

Should we send troops then? It is curious how often we send troops and how seldom it accomplishes much. We should remember Kurt Vonnegut's observation: "I always forget that wars are fought by children." If sending troops is what Christopher Fry called an "Abysmal paucity of imagination," what should be done?

Let's take a look at our objective. There are 18 million Iraqi people, half of them under the age of fifteen, with a pre-war life expectancy of fifty-seven or fifty-eight. About half the adults are illiterate. Half the population works in agriculture, compared to the United States' 2 percent. Before U.N. Resolution 661 in August of 1990, Iraq's oil revenue helped put the population in the mid-range of per capita income, $2,410. Before the destruction of the Gulf War and the hyperinflation that followed, the amount of electricity used per person per year was 750 kilowatt hours compared to our 10,000. They are now down to about 250 kilowatt hours. The country has 7,400 physicians, who have an average patient load of 2,450, nearly five times the load in the United States. And they're having a hard time. Resolution 661 stopped oil exports, froze all Iraqi foreign assets, and prohibited imports, with only food and medicine exempt.

Many of us retain the dramatic picture of the dashing T. E. Lawrence charging across what is now Iraq to capture Damascus in 1917. Not many of us remember that the Ottoman Turks' revenge was a food embargo causing the deaths of three hundred thousand Arabs by starvation. Now this is being repeated in Iraq. More than one hundred thousand children have died from waterborne diseases. That catastrophic public health emergency continues.

This is the cradle of civilization, gently rocked by the Tigris and Euphrates Rivers. A funny thing happened to the Tigris on its way to the sea; it no longer gets there, it no longer has a defined mouth, it just disappears into the swamps. Like the Everglades or the Colorado River, the waters of the Tigris have been siphoned off for irrigation. The resulting salination of farm lands is a terrible problem. The Euphrates is in better shape and makes it all the way to the Shatt al Arab, thus allowing cargoes into Basra. But destruction of sewage treatment plants and electrical generating capacity has made both rivers a mess. By next spring the two rivers, which are the source of drinking water for most Iraqis, are likely to cause thousands more deaths from waterborne diseases, which the woefully underequipped doctors will be powerless to prevent.

Perhaps it's time to recall the Marshall Plan. Getting an honorary degree at Harvard's commencement in 1947, George Marshall predicted an economic collapse in Europe after a hard winter. He feared that the primary benefactor from such chaos would be the Communist Party and proposed the economic aid program that took his name. I looked it up: it cost $17 billion and went to Austria, Belgium, Denmark, Ireland, Iceland, France, Greece, Italy, Luxembourg, Holland, Norway, Portugal, Sweden, Switzerland, Turkey, the United Kingdom, and West Germany. After Truman beat Dewey for the presidency in 1948, he emphasized Point IV of the Plan, described by Harry as, "the purchase of cultural and political loyalty." It became a rule that if the Soviet Union offered aid, we would offer more, and the receiving countries quickly caught on. The Marshall Plan's objective was to rebuild industry and it worked.

But we are thinking of waterborne diseases and Iraqi children. Spring is just around the corner in Iraq, with gastroenteritis and typhoid followed by kwashiorkor and marasmus, the top-of-the-line illnesses from protein/calorie deprivation.

So I set out to measure the problem. I called L. L. Bean and quickly became buddies with Joann, who gave me the specifications on their top-of-the-line water purifier, the one that filters 500 gallons of water down to .005 microns each day. It weighs twenty ounces and twenty purifiers occupy a cubic foot. Retail cost with the first filter cartridge: $139. Then I called a purchasing executive at Bean's. After explanation and wheedling, I got the manufacturer's discount price: $60. Then I figured I would need 66,666 of the purifiers to provide two gallons of pure water every day to everybody in Iraq who needs it. While I was running the numbers, I called the head office of Veterans for Peace in Portland to learn the cost of taking one Iraqi life during the first Gulf War: $400,000. Inflation and technology costs has upped that figure from the $100,000 it cost during the Vietnam War.

So, based on the cost it took to take twelve Iraqi lives in 1991, we have a $5-million opportunity. Four million dollars would buy us 66,666 water purifiers and cartridges, assuming we couldn't get a better price out of the manufacturer than L. L. Bean could. The purifiers would take up 3,333 cubic feet. A 40-foot shipping trailer holds 2,560 cubic feet. So using two containers, transported by sea from Boston to Basra by Hapag Lloyd shipping for $3,800 each, would leave us 1,700 cubic feet for extra filter cartridges and some spare change. As it turns out, gasoline in Iraq is cheaper than water so distribution of our purifiers up the two river valleys via automobile seems feasible.

There is a small problem. Saddam Hussein and his elite are

not very nice people. So one approach to solving the water quality problem would be to take Ralph Waldo Emerson's dictum that there is no limit to what can be accomplished if it doesn't matter who gets credit for it, and let Hussein take the credit. Even if their newspaper readership is only twenty-three people per thousand, word of who sent the purifiers will get around.

The matter of clean water for Iraqi children is of great urgency. The notion of a new foreign policy for the United States--don't send troops, don't threaten countries with aircraft carriers, instead send people who know what to do, how to provide clean water, treat sewage, restore electricity, to generally do well what needs to be done—may take a bit longer. The paradigm for this change is the amazing nine-year "negotiated revolution" in South Africa, when a 300 year old policy of Apartheid was cancelled. As Allister Sparks said in a recent *New Yorker* article, South African society collectively stared into the abyss and having done so, were prepared to recognize their mutual dependence. Collectively we too stare into the abyss, in which lies the mass grave of thousands of children, and we must recognize our mutual dependence, whether from sympathy or from the sheer practicality I referred to in my opening statement.

Perhaps some of you share my impulse to write to Senator Jesse Helms, soon-to-be chair of the Foreign Relations Committee, and say, "From the Book of Judges in the Bible one learns that Samson slew three thousand Philistines with the jawbone of an ass. If you will send me your lower maxillary...."

When we acknowledge diversity and marshal our efforts to preserve it we become a more effective people. And, by taking the greater risk, we become, not incidentally, the safest country in the world.

Virginia Woolf provides my final line:

I meant to write about death,
Only life came breaking in as usual.

A Sense of Wonder: Encounters with Ed Myers

W. DONALD HUDSON, JR.

ED MYERS WAS STANDING on the float, khaki pants tucked into black rubber boots, wearing a crisp white shirt, blue bow tie, and a weather-beaten narrow-brimmed hat set slightly askew on the top of his head. "What do you want to know about mussel farming?"

It was near the end of a fall day nearly thirty years ago, and I had come with a small group of students to learn about aquaculture. Ed held his pipe in his hand as he addressed my students. Questions came slowly, and Ed encouraged more by his careful answers. When asked where he acquired the seed mussels for the long mesh bags in which masses of blue mussels now grew, Ed described with delight how he had made a pact with Mother Nature. His face was alight as his gaze moved from student to student. With just enough theatrical pause—just long enough to catch the students leaning in for the answer—Ed offered an explanation.

Why, all he had to do was suspend ropes in the water column in the spring of the year and within several weeks they were encrusted with tiny blue mussels. Floating in the water, rising and falling on the slightest currents, tiny mussel larvae latched onto the ropes with their first sticky bissel threads. Safe and secure, they spun more threads and there began the quick transformation to the adult form of this common edible bivalve.

Ed's eyes sparkled as he described the bountiful plankton, "mostly diatoms, little golden-brown plants in glass houses," upon which the tiny mussels grew. "You know about primary production, about the food chain, don't you?" We moved from the dock to a small gear shed on the pier, and Ed stuffed a couple of yards of mesh bag material into each student's hand as a token of his or her visit. "With a little effort, I'll bet you could get Mother Nature to work for you in Montsweag Creek."

The memory of my first visit with Ed Myers is as clear today as it was on that late October afternoon. What great teaching and great learning happened in those forty minutes! I can't credit life-changing decisions entirely to those moments, yet I know that each of those students went away from the visit with stronger convictions and renewed confidence in his or her choice to follow paths in the natural sciences. We had mesh bags brimming with tiny blue mussels hanging from a buoy in Montsweag Creek within days of our visit with Ed.

Appreciation of nature was clearly evident on Ed's face and in his voice. We had no difficulty in understanding his love for the little patch of the Damariscotta River at Clark's Cove. Here was a grown man, with grown children, who had not lost his sense of wonder for the complex living world that flowed past his pier. He expressed awe and he marveled at the seeming endless bounty of that wonderful river.

The years passed, and every time I encountered Ed—at a meeting in Augusta, in a shop, or along the street in town—he never failed to encourage good teaching and learning. I always left those chance encounters with a smile, knowing such a great fan of my life's work existed. We always spoke of the sightings of this bird or another, or about a storm just passed, or the report of some commission about the state of our forests. He celebrated the study and appreciation of nature as an essential piece of the education of a child, and the lifelong learning of us all. The Damariscotta River and the mysterious and magical life of the blue mussel had touched Ed's soul, but these were not the only things about which he was passionate. He was a champion for clean air and clean water, as well as for careful management of fisheries--both wild and farmed.

Ed and I spoke at length about climate change and about its influence on the living world when we met at a symposium in May 2002. We were there to learn about the Earth Charter, a global blueprint for sustainable living. Stooped by the years, Ed grasped my hand and looked straight into my eyes. "Wonderful ideas, great discussion! People need to understand the impacts our lives are having on the planet."

Ed and I never spoke again after that day.

I imagine the students who met with Ed Myers on that float in the Damariscotta River nearly thirty years ago remember the day with fondness. They have grown up, with families and careers of their own. One is a documentary filmmaker of the sciences, another a "green" certified forester, and another a teacher. Just like the fine and tenacious bissel threads of the blue mussel, Ed Myers helped them to make a strong connection to the natural world. They have a sense of their place in nature and the world similar, I am sure, to Ed's relationship with the plankton, mussels, and the river.

We cannot underestimate the power of wonder to fuel our spirit and our imagination. Ed Myers lived in this world and he loved life and he never lost a sense of wonder. He taught us to pay attention to the little things that go unseen in the water and the soil and the atmosphere. We should never forget that a great conservationist like Ed Myers was first and foremost a lover of nature.

DON HUDSON is a naturalist and science teacher, and president of the Chewonki Foundation in Wiscasset, Maine.

LETTERS

EDWARD MYERS never was one to stand on ceremony. Whether you were a local Bristol schoolchild or politician, Ed was quite willing to speak his thoughts to you, in print when the occasion merited. He was an educated man yet, perhaps more importantly, he was a human being concerned not only with his own welfare, but with the welfare of his grandchildren and of everyone's grandchildren.

There is neither time nor space to do justice to Ed's extensive correspondence; the following few letters will have to suffice.

[In 2001 the Maine Department of Marine Resources enacted a new regulation requiring seaweed harvesters to apply for and receive a $50 license.]

May 23, 2001

George Lapointe, Commissioner
Department of Marine Resources
Augusta, Maine 04333

In re: Paul Lauzier, Lubec

Dear George:
Some of the papers and some conversations with my son Allen's fellow Quakers about Paul Lauzier and gathering seaweed have been vouchsafed to me in recent days.

My view of this "case" is that the man should be canonized and pollinated into a mentor for the seaweed "industry," not threatened with arrest and jail time for taking a principled stand.

This man has apparently expended a decade and a half in making a carefully sustainable effort in harvesting a half-acre of rockweed in order to make an equally sustainable home garden which has provided him with his year's needs of vegetables. He does this, year after year, on foot and in a skiff under his own renewable and non-polluting power.

What this young man (when you're in your eighty-fifth year, everybody is young) is doing reflects the subsistence way of life which brought Maine practically unscathed through the Great Depression seventy years ago; it is also a rare harbinger of overcoming "the terrors that are to come."

He and the few others like him do not fit the laws and regulations designed for mechanical (and polluting) harvest; the fifty-buck fee seems to be nothing more than a protection to the big guys. Lauzier and colleagues should be cherished, not punished.

We gathered alternate European oysters which had set naturally on mud bottom, measured both sets, and placed one set on hard bottom on our shore. Our intent was to measure comparative growth rates. One day a large rockweed barge came to our shore with its knives going. Neither captain nor crew would acknowledge our hails from shore. So we went alongside in our work scow. They would not acknowledge our presence, and we had some fear of being shot or run down. They gathered their ton in a net bag and took off. Our oyster experiment was utterly destroyed. This took place within an aquaculture leasehold, God wot. In our opinion, DMR's regulation don't amount to a peehole in the snow.

Please write Paul Lauzier and lift this departmental incubus from his shoulders. He deserves support, not harassment. If nothing else, "De minimis non curat lex." And while you're at it, hire him for your calligraphy needs—he's good at it.

Yours faithfully,

May 4, 1999

DEAR NEIGHBORS:

If you judge by the wondrous way Elinor Edlund is tending the spotlights on the corner of her house, my *Working Waterfront* column on lights from Yale's observatory in Chile up to the Arctic northern ones must have had some local circulation. As we watch the moon and the clouds through the skylights of the "heart attack" ground-floor bedroom, Julia and I really appreciate her thoughtfulness.

The town warrant, Articlc 5, about the review of street lights, should be on the agenda of the meeting of the selectmen at 7:00 P.M. on Thursday, May 6, at the new townhouse.

I expect to be there. Since there may be some varying opinions, I thought you should know about it.

The column about light pollution has, somewhat to my surprise, brought forth more favorable feedback than any of the hundred or so others I have written—from three states. One guy in southern Maine is very upset about a 200- to 300-unit retirement home planned directly opposite his farm across a small river. He figures he'll never see a complete sunset again.

Yrs faithfully,

December 30, 1998

President William J. Clinton
The White House, Washington

My dear Mr. President:

I presume to advise you on your procedure in the matter of impeachment as follows:

1. The President is Commander-in-Chief;
2. As that, he has a commission at least equal to field grade or higher in the armed forces;
3. All commissioned officers are gentlemen, by Act of Congress;
4. Gentlemen do not discuss their love lives, whether or not under subpoena or oath;
5. Period.

If any question arises about the skein of law, you may send them to Hookham Frere's translation of Aristophanes *The Frogs,* published in 405 B.C.E., about the third page:

> Talk about...a person's soul...not being perjured
> When...the tongue...forswears itself...in spite of the soul.
> 'Men's fancies are their own—Let mine alone'

(The insertions of...are in the original. This is therefore a complete quotation, and will serve as earliest.)

After your declining to testify on this subject, you will be held in contempt of Congress, which will allow you to join the rest of us; your popularity rating will rise fifteen points.

Also rising shortly thereafter will be the members of the Senate, who will vote 55–45 or anyway less than 60–40, then fuss about censure, and then subside.

Starr was hired to probe Whitewater and succeeded: culprits were brought to justice; you and your spouse completely absolved. But then Starr went on to become a voyeur, the lowest form of sexual deviant. This is his tragedy, not yours, and if he merits anything it's pity.

Follow the above; then go back to work for the next two years doing constructive things which are crying to be done.

And salute Hillary.

Yours faithfully,

October 11, 1998

DEAR ROBIN [Alden, then Department of Marine Resources commissioner]:

That was a perfectly astonishing and immensely gratifying performance you and Mike put on after lunch on Friday. As you know, I am somewhat deaf, and it was not until I heard "bow tie" that I had the first inkle that my life was about to be changed. Wondrously.

I would rather have the Foghorn than the Nobel.

Where on earth did you remember my use of "coruscant failure" as self-description? I think that coruscant means glittering, rather than brilliant, the latter being more than an invasion of modesty.

It is almost scary what a wonderful time we have all had these last few decades, fighting Maine Yankee and NMFS, coming to careful compromise with ASMFC, dancing gingerly ballets with the federals, never giving up on the hopeless task of trying to knock some sense into the legislature. I had the added fascination of following your career; being commissioner and acquiring part ownership of the Ship of State—like owning a yacht, there are two perfect days: The day you take it on and the day you get rid of it. What a job you did in-between!

Robert Frost at the end of "Two Tramps in Mudtime":

Yield who will to their separation,
My object in living is to unite
My avocation with my vocation,
As my two eyes are one in sight.
Only when love and need are one,
When the work is play for mortal stakes,
Is the deed ever really done
For heaven's and the future's sakes.

I've used it a lot on one podium or another, but this is the first time anyone thought to apply it to me. I thank you from the bottom of my cussed heart.

Love,

July 18, 1998

Dr. Richard Grossman
S. Yarmouth, Maine

Dear Richard:

"Engage us." "Conspire with us." Who, me?

On June 27 last, I lectured the Camp Kieve Mother and Daughter Science Canoe Trip when it stopped by this wharf in a driving rain. I softened them up by setting the executive suite thermostat to 85°F and allowing them to make free with my winter clothes rack (one of the ladies put on a raccoon coat acquired sixty-five years ago at a small college that was into such garb, worn by me in breaking ice around our aquaculture gear, only no longer because climate change has brought no ice to this estuary for five consecutive years), plus a giant plate of hot mussels with horseradish sauce to warm their insides, and then invited them to get furiously angry:

> All you mothers nursed all you daughters and never gave it a thought, except that it was the right and healthiest thing to do—and it doubtless was. BUT, if any of you had a baby last year in the Netherlands, supposedly one of the most civilized countries in the world, and you breast-fed her, she'd have had plasma PCBs 360 percent higher than formula-fed babies. PCBs are polychlorinatedbiphenyls, from the manufacture of electrical equipment, discarded batteries, and fumes from the factory smokestacks that made them. AND the baby would have absorbed in her first months fifty times the dioxin that we're exposed to daily as adults. What you do for openers is shout a

> Latin phrase that company lawyers will recognize (and I led the group in a unison shout): *Quo warranto!* By what authority, by what right, do you poison me and my baby? Is this really what you wanted to do when you grew up? Don't you have children? Have you ever seen anything more beautiful than a mother nursing—it's the essence of the creativity and continuity of life—of nurturing from generation to generation—and you poison this to make a buck on alkaline batteries for video games and speck-free rolls of paper for cash registers? Can you be serious?
>
> I once took a ride on a mussel-delivery truck from Zeeland to Brussels; if a Dutch mother had ridden with us, she would have been committing a crime—carrying containers of dioxin and PCBs over the legal limit across an international border. How absurd can you get? Most of you have phone cards. When you get back on the mainland, if you know an executive in Phillips NV or Shell or the Energizer Bunny or Big Macs or Haagen-Daz, call him up. If you happen to know a Dutch one, roll out at four in the morning and make a call; they know what time it is where you are, and you can really get their attention.

And so on.

On June 28 last, I conducted the service at the Unitarian Universalist Church in this area with a sermon title of "The Embodiment Impolitic," an essay which drew heavily on (and with full credit to) Messrs. Grossman and Kellman. The old school chum who hired me called it superb, which I can convey cheerily as it's testimony to the excellence of Kellman's history of the Maine Corporation. The peroration was much the same as the Kieve ladies, in urging them to writefaxphone execs they knew in Maine or anywhere. By a fortunate coincidence, Maine shrimp was among the refreshments and I had already explained the levels of dioxin in the shrimp from the paper companies to the sea via the St. Croix, the Machias, and the Penobscot and carried west by the Labrador Current. At the end I used Lord Acton's letter to Bishop Creighton, the Nixon Administration's J. R. Haldemann's variation ("Power corrupts and absolute power is really neat") and John Galsworthy's enlargement, which you may find use for:

> Nothing so endangers the fineness of the human heart as the possession of power over others; nothing so corrodes it as the callow or cruel exercise of that power; and the more helpless the creature over whom power is callously or cruelly exercised, the more the human heart is corroded.
>
> That completes my report.

Yrs faithfully,

Correspondence between Edward Myers and Nobel prize-winning author, Professor E. O. Wilson, Harvard University

March 5, 2002

MY DEAR DR. WILSON:

My eye fell on the *Globe*'s paragraph about "super-organisms" as one-third of the planet's biomass, and it raised some questions, all of which have been answered by the wondrous book.

I have been having fun with my occasional audiences by asking them to visualize a line of square miles atop an ocean depth of 5,280 feet and asking them to guess how many cubic miles would be needed to contain all of humanity:

- Average weight (somewhere between a Sumo wrestler and a Thai dancing girl, including the 55 percent of Americans who are overweight): 150 lbs. = 2.5 ft^3
- Times world population of 6.25 billion
- Equals a rounded total of biomass of 16 billion ft^3
- A cubic statute mile equals 147 billion ft^3
- So humanity would occupy 11 percent of the first mile, or 580 feet in depth.

The usual guesses, so far, range from 7 to 212 $miles^3$, from a non-professional audience. The answer makes them thoughtful, amazed that such a small pile of humanity can wreak such havoc and dazzled that 8 or 10 billion still would not fill the first mile.

For a course in climate disruption I shall be giving at the Senior University of Maine in September, I am specifying *The Future of Life* as the text. The text limit is $20, so I assume the University can buy a dozen or two without doing damage to your royalties. The dust jacket and concordance are alone worth the difference.

Perhaps you may not have seen "Letter from a Distant Land," another letter to Thoreau, this one in 1956 from Philip Booth, who was teaching at Wellesley and resident in Lincoln when he wrote it. His present address is Castine, Maine, 04421, which is all the address needed.

Faithfully yours,

I grieve that Wendell Berry has fallen into bitterness and nursed indignation that we can't all become disciples of Thos. Jefferson. We do not have that kind of time. WB must be getting old. Huston Smith's problem is beyond my wildest surmise.

14 March 2002

Dr. Edward A. Myers
PO Box 551
Damariscotta, ME 04543-0551

Dear Dr. Myers:
Thank you for your very interesting letter, and the earlier letter to Thoreau, which I hadn't seen (one other, from T. H. White, has also been called to my attention).

Also, thanks for the calculation on the volume of existing humanity. I've been going around saying 3 cubic miles to make the same point you do, based on a back-of-the envelope calculation I made some years ago. I re-calculated after receiving your letter, with the average person taking up (when log-stacked on top of others) 68" x 12" x 10" or 8,160 cubic inches. The human population of 6.2 billion people, if I haven't blundered somewhere, thus occupies about one-fifth of a cubic mile. So you're right and I was off an order of magnitude. What's the difference between your 0.11 and my now 0.20? Could it be that you squished everybody together into a single blob when converting biomass to volume, while I've just log-stacked and left space for body and appendage curvatures?

Your response eagerly awaited.

Warmly,

Edward O. Wilson

March 18, 2002

DEAR DR. WILSON:
You'd best stick with the back of your envelope.

You are correct that I squnched [*sic*] the biomass on the basis of cubic feet, with no interstitial allowances for protuberances. As a matter of record, my first try was on behalf of my (volunteer) eighth-grade class in South Bristol Elementary School, wherein the scholars are more familiar with the nautical miles, rounding down to 6,000 feet, a cube I can compute in my head at 216 billion ft^3.

With 6.25 billion times 2.5 ft^3, the result is even more dramatic—on my abacus just 7.24 percent of a cubic nautical mile or 434 feet. That such a small glob can cry so much havoc and let slip so many dogs of war.

Next Sunday I am supplying the rostrum of the local Unitarian Universalist Fellowship with a review of *The Future of Life* plus digressions. If I ever get it written, I'll burden you with that.

Thanks for the doctorate. I do have an honorary one from an obscure local university; the citation was its last official act before being submerged in the University of Southern Maine. So I have the three bars on the robe, but when I put it on, I'm all dressed up with nowhere to go.

Yrs faithfully,

Oh, yes, I did do the Archimedes bit with a piece of duct tape holding a rule with one end at unoccupied water level of the tub. 153 lbs. Displacement by a 161-lb. bod. But when I ducked my head, my knees came out, so it was very bucktailed science. Turned out my students all took showers only (in Maine parlance, "I take a shower once a month whether I need it or not"), so they could not prove the positive buoyancy that I did. As Philip Booth said in his letter to Thoreau, "My classes are good failures."

To Peter Shelley, Maine director of the Conservation Law Foundation

May 15, 1998

Dear Peter:

For too many decades I have been wrestling on one board or another with the problem of geographical representation of an equitable quality. In Maine, it cannot be done. A sixteen-member committee, which can be unwieldy, always gets complaints from Aroostook (12 people per square mile) at being neglected by Cumberland (283 people per square mile). And somebody always wants to save Washington County from itself, which resents that, while wishing that Western Maine could be ceded to New Hampshire. I say the hell with it. Put together a committee that will work. When some county heats up, the needed people will appear and be added on or join as consultants or expert witnesses. And donate money.

It is all one project. Maine's share, and it is considerable, of the 6.75 billion tons of carbon particulates annually, 73 percent in the Northern Hemisphere, needs to meet a metabolizing force to take care of it, and there are only two: trees and vegetation on the one hand and the Atlantic Ocean on the other. If you clearcut the trees, pave over the vegetation, and pollute the ocean, you have had it.

The Gulf of Maine, as Henry Bigelow observed seventy-five years ago, is a peculiar piece of water—it needs the 75–100 cubic kilometers of spring runoff from the watersheds beginning in the North Woods to provide the right mixture for the herring on Stellwagen Bank and Georges, with the Labrador Current driving the fresh water to the high-rent district of Hingham, out east to Provincetown, back north to Cape Sable, and around again. If it's laden with dioxin and whatever the paper companies spray with—we used to find DDT in the shrimp Down East—the system won't work, the Atlantic right whale will proceed to extinction, the Gadidae won't recover; the circle will close.

Rockland is the ideal central location for this project. It's at the debouchement of the Penobscot watershed and it's 35 miles from Seguin to Matinicus and the same from Matinicus to Great Duck.

...Let us never forget that Mother Nature bats last.

Yrs faithfully,

June 7, 2001

HON. SUSAN M. COLLINS
172 Russell Office Building
Washington, D.C. 20510

Meditation (of all things)

Dear Senator Collins:

You haven't time for them, but perhaps an intern can be set to reading Gibbon's *Decline and Fall of the Roman Empire* and Carlisle's *French Revolution,* while setting down for you all that is germane in both to the U.S.A.'s present headlong rush to arm space.

The intern's notes (after some weeks of concentration—the two titles usually run to eight volumes) will provide you a perfect prescription for the collapse of the United States. In the case of France, the 1789 revolutionaries gave way to Napoleon. In the case of Rome, over-extension plus a plethora of hubris brought collapse from within. Hannibal and his elephants merely signed the conclusion.

This last year has been a golden moment: The U.S.A. has no national enemies—oh, there are a couple of Rogue States but these or others like them will be around for the foreseeable future; starved North Korea, sanctioned Iraq, embargoed Cuba, unsettled Iran. If we have the right idea, however, why don't we set an example instead of planning new ways to threaten or kill people?

And if we are so wealthy, why don't we pay down the national debt, rather than leaving it for our grandchildren? Dissipating a surplus to the 1 or 2 percent who don't need [it] seems very unwise. It didn't work for Reagan, and it didn't work for the current President-Select. After billions of dollars, we can't seem to interdict more than a random few pounds of cocaine, the scourge of this country.

So get the intern's notes and spend fifteen minutes every day in meditation on the long view. Including how you wish to be remembered by history—as some sort of swing vote during the Dubya administration? Or as someone who learned from history's lessons.

All the high-tech weapons from nuclear submarines to the far reaches of the sky will be useless against the terrorism of hate, when we have earned our enemies and alienated our friends. We'll be fighting anthrax and Ebola and heaven knows what else, while retaliating against innocent people.

If you do nothing else, get your intern to exhume E. B. White's "The Morning of the Day They Did It" from old files of the *New Yorker*. If that doesn't work, get Shelley's sonnet "Ozymandias."

Yours faithfully,

HIGHLIGHTS OF PROGRESS

Highlights of Progress

MELISSA WATERMAN

AS ED QUIPPED IN HIS WRITINGS, he acted as the "local Cassandra without portfolio," a harbinger of possible doom without the credentials of a scientist. While his columns and sermons do highlight the potential disasters that await the world if significant changes are not made in our collective economic and social behaviors, Ed never forgot that the world is full of people making a difference. Whether that difference took place on the local level, as Ed exemplified with his purchase of a hybrid gas-electric vehicle, or at the global level, people are finding new ways to protect the air, water, and soil of our shared earth. And many are making a living doing so. These are real people, making real impacts on the world, neither wild-eyed idealists nor impoverished dreamers. Like Ed, they realize that there are problems, that these problems are *not* unsolvable, and have applied their personal talents and energy toward their solutions. Ultimately people's efforts are the source of change in the behavior of corporations and, just as surely though often not as quickly, in the behavior of governments.

Ed focused time and again on the danger altering the climate poses for ourselves and the natural world. He pointed out the importance of switching our economy from its dependence on fossil fuels to fuels derived from the inextinguishable power of the sun. Though not at the level that Ed wanted and certainly not at the scale the world's future requires, renewable energy sources such as wind power, solar power, and tidal power are being drawn upon today.

Several international corporations are investing their money and technical expertise to make wind power a commercial reality. Shell Renewables and Eskom, South Africa's national electricity supplier, are developing the world's largest solar rural electrification project. The project, located in the Eastern Cape area of South Africa, will bring solar electricity to 50,000 rural homes. To date the Shell-Eskom joint venture has installed 6,000 Powerhouse solar home systems in towns and small communities, bringing electricity to an estimated 30,000 people in the area. The two companies are promoting a utilities business model for use in other countries around the globe. Currently Shell is working with partners in India, Sri Lanka, and the Philippines to create renewable energy systems for rural communities.

In the United States, Shell Renewables created a division called Shell WindEnergy Inc. According to the company, demand for renewable energy continues to grow in this country by 25 percent each year. Shell WindEnergy purchased the Llano Estacado Wind Ranch in Texas, an 80-megawatt wind power facility. Llano Estacado can produce enough power from the wind to meet the needs of 30,000 homes. Shell WindEnergy also owns and operates the 50-megawatt Rock River facility in Wyoming.

Cielo Wind Power LLC operates the world's largest wind energy plant, a 278-megawatt facility in Texas. Overall, the company produces approximately 600 megawatts of electricity from wind each year. Elsewhere in the United States, Mid-American Energy Company plans to build a huge wind farm in Iowa. Funded by investor Warren Buffett, the farm will feature 200 wind turbines, capable of generating 310 megawatts of electricity. When complete, the project will be capable of providing electricity to 85,000 homes.

Utah's largest power company is promoting commercial use of renewable energy through its Blue Sky Program. Utah Power Inc. started the Blue Sky Program in 2000 to allow customers to choose energy based on its source and to foster a demand for renewable energy. The company offers customers the opportunity to buy renewable energy in 100 kilowatt increments for an additional $2.95 per 100 kilowatts. Uinta Brewing Company of

Salt Lake City took up the company's offer and built a whole new brewery and pub run entirely on wind power. According to the brewery's president, 357,120 fewer pounds of carbon dioxide will be released into the atmosphere because of the use of wind power.

Solar power is another renewable energy source that has come a long way in the past two decades. In the United States and abroad, private companies and governments are pursuing commercial applications of solar power at a scale unthinkable a generation ago.

Australia has mandated that electricity companies within the country supply 9,500 gigawatt hours of renewable energy a year by 2010. In response to that requirement, EnviroMission Ltd. plans to build a 3,300-foot solar tower in New South Wales. The 200-megawatt solar tower will be about as wide as a football field and will stand in the center of a 4.3 mile-diameter roof of glass. Energy will be generated when the sun heats the air under the upward sloping glass roof. As the hot air rises, the tower itself creates a powerful updraft. That updraft pulls the air through the tower, where it is used to spin 32 turbines, 24 hours a day. The solar tower was invented by a German engineering firm, which constructed a 656-foot-high demonstration power plant in 1982 in Spain. That 50-kilowatt plant produced electricity for seven years. EnviroMission says that by 2006 the solar tower will generate enough electricity to supply 200,000 homes. As a result, 700,000 fewer tons of greenhouse gases will be released into the atmosphere.

Solar energy projects are not all at such a grand scale. The international energy company, BP Solar, and the governments of Spain and the Philippines signed an agreement to bring solar power to about 400,000 residents in 150 villages of the Philippines. The $48 million contract is financed by the Spanish government and is the largest solar energy project ever undertaken by BP Solar. The first phase of the project will bring solar panels to 5,500 homes in 70 villages. The panels will power household lights, 25 village irrigation systems, and 97 drinking water distribution systems. In addition, solar panels will allow 68 schools and 35 health clinics to have electrical lights for the first time. As a result, less wood will be cut in the area to use as fuel, reducing deforestation and the emission of carbon into the atmosphere.

Back in the United States, the Rancho Seco nuclear plant is now a fully functioning solar energy facility. In 1989, local residents voted to shut down Rancho Seco. After much study, they found that by using renewable energy sources only, they could have the same amount of electricity at a comparable cost. The dormant nuclear plant, owned by the Sacramento Municipal Utility District, is now surrounded by vast fields of photovoltaic panels which generate approximately 4 megawatts of electricity. The area's 533,000 customers receive their electricity from a variety of sources: 57.4 percent from hydroelectric; 41.5 percent from thermal power, 0.7 percent from photovoltaic panels, and 0.4 percent from wind power. Wringing power from any and all sources, the Utility District recently purchased a landfill-gas generator system, which extracts methane gas from the county landfill to generate electricity.

Even the White House has begun to recognize solar power's benefits. A 9-kilowatt photovoltaic system was installed in January 2003 on the roof of a White House grounds building for the National Park Service. The photovoltaic system feeds solar-generated power directly into the White House grounds distribution system, providing electricity wherever it is needed. Two solar thermal systems, one to heat the pool and spa and one to provide domestic hot water, were also installed.

Tidal power also is receiving attention from companies around the world. Traditional tidal dams are still operating at La Rance, in France and in Annapolis, Nova Scotia. These dams trap water at the peak of high tide. Vents at the bottom of the dam release the water past turbines that turn to generate electricity. La Rance produces 240 megawatts and the Annapolis facility produces 20 megawatts of electricity.

In 2003 a new tidal power system was inaugurated off the coast of England. An undersea turbine located on the seafloor about one mile off Devon can produce 300 kilowatts of electricity. The turbine uses a single 11-meter-long rotor blade, like that of a windmill, to produce power as the tide comes in and again as the tide goes out. According to the British manufacturers, approximately 10 gigawatts (10 billion watts) of power could be drawn from tidal currents in English waters. Undersea turbines such as this exploit the strong tidal currents which are often found in shallow ocean areas, particularly near headlands or islands. The tidal current turbines sit anchored to the seabed and have the ability to pivot to face into the changing current.

In addition to the English project, one 5-kilowatt Japanese turbine has been running since 1990 and another, at the tip of Norway, recently began generating approximately 300 kilowatts of electricity to the small town of Hammerfest. Another test installation of undersea "windmills" operates in a remote area of the Amazon River in Brazil. There, local residents use the turbines to recharge dozens of car batteries to run their television sets. Supporters of these undersea turbines believe the structures will be more acceptable than wind power because they are less obtrusive and because tidal currents are much more dependable than the wind.

As Ed Myers realized fully, individuals around the globe are striving every day to make the world a better place. You may not read about them in the daily newspaper or hear much about their explorations on the television. Yet they exist and are moving the world forward toward a new future every day. The question Ed posed time and again is: When will we, the people, convince governments and corporations to convert to sustainable use of the earth's resources in order to prevent irredeemable harm to ourselves and our planet? Can you and I change our own behavior in order to live within the world and not apart from it? It is up to every one of us to use our intelligence, technology, and spirit to, in the words of Edward Myers, "do it right this time."

MELISSA WATERMAN is an independent writer living in Rockland, Maine. Her articles on marine environmental topics have been published in *National Fisherman, Coastal Living, Conservation Matters,* and other national publications. Ms. Waterman counts her friendship with Edward Myers as a highlight of her experiences in Maine.

BIBLIOGRAPHY

Bibliography

(Allen Myers helped with early research on this project.)

Alexander, James and Stanley Nider Katz, eds. *A Brief Narrative of the Case and Trial of John Peter Zenger, Printer of the* New York Weekly Journal. Cambridge: Belknap Press, Harvard University Press, 1972.

Alley, Richard B. *The Two-Mile Time Machine: Ice Cores, Abrupt Climate Change, and Our Future.* Princeton: Princeton University Press, 2000.

Anderson, Ray C. *Mid-Course Correction: Toward Sustainable Enterprise: The Interface Model.* Atlanta: Peregrinzilla Press, 1998.

Atkinsson, Alan. *Believing Cassandra: An Optimist Looks at a Pessimist's World.* Vermont: Chelsea Green Publishing Co., 1999.

Ayres, Ed. *God's Last Offer: Negotiating for a Sustainable Future.* New York: Four Walls, Eight Windows, 2000.

Barzun, Jacques. *From Dawn to Decadence, 1500 to 2000.* New York: HarperCollins Perennial, 2001.

Benyus, Janine M. *Biomimicry: Innovation Inspired by Nature.* New York: William Morrow and Co., Inc., 1997.

Berry, Wendell. *Collected Poems, 1957–1982.* San Francisco: North Point Press, 1985.

_____. *The Country of Marriage.* New York: Harcourt Brace and Co.

_____. *Sex, Economy, Freedom, and Community.* New York: Pantheon Books, Random House, Inc., 1993.

Bigelow, Henry B. and William C. Schroeder. "Fishes of the Gulf of Maine," *Fisheries Bulletin #74, Vol. 53.* Washington, D.C.: GPO, 1953. Reissued by Woods Hole Oceanographic Institute and Harvard Museum of Comparative Zoology, 1964.

Blunt, Edmund and George Blunt. *The American Coast Pilot.* 14th ed. New York, 1837.

Bly, Robert. *The Sibling Society.* New York: Vintage Books, Random House, Inc., 1996.

Bright, Chris. *Life Out of Bounds: Bioinvasion in a Borderless World.* New York: W. W. Norton and Co., Inc., 1998.

Brown, Lester R. *Eco-Economy: Building an Economy for the Earth.* New York: Earth Policy Institute, W. W. Norton and Co., Inc., 2001.

Chappell, Tom. *Managing Upside Down.* New York: William Morrow and Co., Inc., 1999.

Colborn, Theo, Dianne Dumanoski, and John Peterson Myers. *Our Stolen Future.* New York: Penguin Books USA, Inc., 1997.

Coles, Robert. *The Secular Mind.* Princeton: Princeton University Press, 1999.

Cruttwell, Peter. *History Out of Control: Confronting Global Anarchy.* Devon, UK: Green Books Ltd., 1995.

Durning, Alan. *How Much Is Enough? The Consumer Society and the Future of the Earth.* New York: W. W. Norton and Co., Inc., 1992.

Eisely, Loren. *The Invisible Pyramid.* New York: Scribner's Sons, 1970.

Falk, Richard A. *A Global Approach to National Policy.* Cambridge: Harvard University Press, 1975.

Friedman, Thomas L. *The Lexus and the Olive Tree: Understanding Globalization.* New York: Anchor Books, Random House, Inc., 2000.

George, James. *Asking for the Earth, Waking Up to the Spiritual/Ecological Crisis.* Rockport, MA: Element, Inc., 1995.

Hay, John. *The Run.* Rev. ed. Garden City, NJ: Anchor Books, Doubleday and Co., Inc., 1965.

Holy Bible, The. King James Version. Philadelphia: A. J. Holman Co.

Holy Bible Concordance. Cleveland: World Publishing Co., 1962.

Horsley, Richard A. and Neil Asher Silberman. *The Message and the Kingdom.* New York: Grosset/Putnam, 1997.

Jackson, Wes, Wendell Berry, and Bruce Coleman, eds. *Meeting the Expectations of the Land.* San Francisco: North Point Press, 1984.

Koch, Maryjo. *Pond Lake River Sea.* San Francisco: Swans Island Books, Collins Publishers, 1994.

Loewen, James. *Lies My Teacher Told Me: Everything Your American History Book Got Wrong.* New York: New Press, 1995.

Mansfield, Howard. *The Same Ax, Twice: Restoration and Renewal in a Throwaway Age.* Hanover, NH: University Press of New England, 2000.

McDaniel, Carl and John M. Gowdy. *Paradise for Sale: A Parable of Nature.* Berkeley: University of California Press, 2000.

Milne, Lorus J. and Margery Milne. "Will the Environment Defeat Mankind?" *Harvard* (Jan.-Feb. 1979): 19-23.

Miner, R. W. *Field Book of Seashore Life.* 8th ed. New York: G. P. Putnam's Sons, 1950.

Moore, Hilary B. *Marine Ecology.* New York and London: John Wiley and Sons, Inc., 1958.

Morris, Henry M. *Science and the Bible.* Rev. ed. Chicago: Moody Press, 1986.

Orr, David W. *Earth in Mind: Our Education, Environment, and the Human Prospect.* Washington, D.C.: Island Press, 1994.

Outwater, Alice. *Water: A Natural History.* New York: HarperCollins, 1996.

Oxford English Dictionary, The. Compact ed. London: Oxford University Press, 1971.

Perrin, Noel. *Life with an Electric Car.* New York: W. W. Norton and Co., Inc., 1994.

Putnam, Robert D. *Bowling Alone: The Collapse and Revival of American Community.* New York: Simon and Schuster, 2000.

Quinn, Daniel. *My Ishmael: A Sequel.* New York: Bantam Books, 1997.

Ray, Jannisse. *Ecology of a Cracker Childhood.* Minneapolis: Milkweed Editions, 1999.

Regier, Henry A. *A Balanced Science of Renewable Resource, with Particular Reference to Fisheries.* Seattle: University of Washington Press, Washington Sea Grant, 1976.

Reiss, Robert. *The Coming Storm: Extreme Weather and Our Terrifying Future.* New York: Hyperion, 2001.

Schumacher, E. F., *This I Believe and Other Essays.* Devon, UK: Resurgence Books, Green Books Ltd., 1997.

Shlain, Leonard. *The Alphabet Versus the Goddess: The Conflict Between Word and Image.* New York: Penguin Group, 1998.

Smith, Huston. *Forgotten Truth: The Primordial Tradition.* New York: Harper & Row Pub., Inc., 1976.

Stilgoe, John R. *Alongshore.* New Haven: Yale University Press, 1994.

Thucydides. *History.* Ann Arbor: J. W. Edwards Pub.

Tickell, Joshua and Kaia Tickell. *From the Fryer to the Fuel Tank: The Complete Guide to Using Vegetable Oil as an Alternative Fuel.* 2nd ed. Sarasota, FL: Green Teach Publishing, 1999.

Trager, James. *The People's Chronology.* Rev. ed. New York: Henry Holt and Co., 1992.

Tuchman, Barbara. *The March of Folly: From Troy to Vietnam.* New York: Alfred A. Knopf, 1984.

Vernadsky, Vladimir I. *The Biosphere.* New York: Springer-Verlag, 1998.

Ward, Barbara. *The Rich Nations and the Poor Nations.* New

York: W. W. Norton and Co., Inc., 1962.

Wheatley, Margaret and Myron Kellner-Rogers. *A Simpler Way*. San Francisco: Berret-Koehler Publishers, Inc., 1996.

Wilson, Edward O. *Consilience: The Unity of Knowledge*. New York: Vintage Books, Random House, Inc., 1999.

____. *The Future of Life*. New York: Alfred A. Knopf, 2002.

Worldwatch Institute. *State of the World 2002*. New York: W. W. Norton and Co., Inc., 2002.

Since 1926, when Kieve Affective Education began as a small boys' camp on the shores of Damariscotta Lake, it has been the organization's belief that everybody on this earth has an equal right to inherit the world in as good or better state than their predecessors found it. With that right goes an obligation: to ensure the integrity of what we have all been given.

Kieve Affective Education, through its Science and Wilderness Programs for Girls and Women, Ocean Term, Leadership Decisions Institute, and other programs for children and adults, promotes its mission to empower young people and the adults who affect them to contribute positively to society. Over the decades Kieve has helped thousands of people develop the skills to know the natural world and to explore their personal relationship with it.

Richard D. Kennedy, Founder
Kieve Affective Education, Inc.

Kieve Affective Education, Inc.
PO Box 169
Nobleboro, ME 04555
(207) 563-5172 • www.kieve.org